MW00581766

Implementing the ISO/IEC 27001 ISMS Standard

Second Edition

For a complete listing of titles in the
Artech House Information Security and Privacy Series
turn to the back of this book.

Implementing the ISO/IEC 27001 ISMS Standard

Second Edition

Edward Humphreys

Library of Congress Cataloging-in-Publication Data
A catalog record for this book is available from the U.S. Library of Congress.

British Library Cataloguing in Publication Data
A catalogue record for this book is available from the British Library.

Cover design by John Gomes

ISBN 13: 978-1-60807-930-8

685 Canton Street
Norwood, MA 02062

10 9 8 7 6 5 4 3 2 1

This book is dedicated to my father (Thomas Humphreys)
and to my mother (Alice Humphreys),
to my sons (Alexander, Thomas and James),
to my grandchildren (Siena, Juliet, Penelope and Georgia)
and last but not least to Angela Elizabeth

Contents

Acknowledgments

I would like to thank all those reviewers for their suggestions and invaluable comments. I would also like to thank my colleagues and friends in the JTC 1/SC 27 for their supportive suggestions.

Introduction

Information security is a global issue affecting international trading, electronic commerce, mobile communications, social media and the various systems and services that make our digital world and national infrastructures. Managing information security is an even more critical issue, as it involves using and managing the policies, procedures, processes, control measures and supporting applications, services and technologies that are needed to protect the information that organisations, governments, consumers and citizens rely on to carry out their business and to live their lives. Information security management needs to be effective, suitable and appropriate if it is to protect information from the risks that business and society faces in this digital age. For example, processing information is essential for those systems and services we rely on to provide transportation, telecommunications, health care, the supply of sources of energy (such as electricity, water and gas) and many other common things we accept as being there when we want them. The information systems that provide vital management support for these different aspects of this public infrastructure are vulnerable to a wide range of information security risks.

Our dependency on digital information is all pervasive, as are the risks to this information. Information could be disclosed and accessible to unauthorized users, corrupted or modified either in some unauthorised or accidental way, or lost or unavailable due to a system failure. There are three important security objectives mentioned here—protecting the confidentiality, integrity and availability of information. Understanding what the risks are and assessing how these risks affect and impact business in terms of these objectives is central to being able to manage these risks effectively.

An organisation needs to assess its risks in terms of the potential impact that a security incident might have on its business and the likelihood of this security incident occurring. It needs to adopt an approach to risk assessment that is effective, suitable and appropriate to its business. The simple message to all organisations and governments is that investing in the management of information security risk is wise.

All organisations are dependent more than ever on information and ICT for driving their business processes, productivity and efficiency, and so the need to protect this information is vital to their survival in today's digitally driven economy. Assuring the confidentiality, integrity and availability of this information, as well as protecting personal data, is at the core of business success.

This book covers some of the important international standards on information security that will help organisations to protect their information. These standards (the so-called ISO/IEC 27000 ISMS Series) were developed over a number of years to meet the needs of business and commercial markets and to protect their information, and to prosper in today's dynamic business world.

This book covers what is contained in these standards as well as practical advice on their implementation. A number of use cases are provided to illustrate the use of these standards in business situations. In particular, it includes what is involved in establishing and designing, implementing and deploying, monitoring and reviewing, and updating and improving an information security management system (ISMS). Through the use of a well-tried tested process model that is embedded in ISO/IEC 27001, the key ISMS standard enables a business to establish an effective information security regime and risk management process and to deploy a number of management controls to continually improve the ISMS to ensure that they still have effective information security. The process of third-party certification is covered, which enables business to demonstrate through independent audits that they have in place an information security management system that effectively protects their information.

The author of this book cannot be held legally responsible or in any way liable for the real world use, application, implementation, or interpertation of the information given in this book. Users and readers alone are wholly responsible for the correct use, application, implementation, or interpertation of the information given in this book.

Finally, the spellings of words used in this book are the internationally accepted spellings adopted and used in ISO and ISO/IEC standards and are from the *Oxford English Dictionary* (OED).

CHAPTER

Contents

Information Security

1.1 The Importance of Being Informed

Do nothing secretly; for time sees and hears all things, and discloses all.

—Accredited to Socrates

Historically information[1] has always played an important role, from ancient times to the present day. For example:

- Its use in the military campaigns of Alexander III of Macedon, Xerxes, Julius Caesar and others; its use in managing the societies and cultures of ancient Greece, Egypt, Persia and China and across ancient cultures and its use to help the Phoenicians, Greeks, Persians and Romans trade across their empires;
- Its role shaping our views and thoughts of the world through progressing our knowledge and understanding of mathematics, science, medicine and philosophy from ancient times to the modern era;

1. Informare (to give form to the mind), πληροφορια from πληρης (fully conveys).

- The silk and spice routes, involving countries in Europe and the near, middle and far east, trade and navigation of the Renaissance merchants and seafarers all were dependent on the availability and accuracy of information;
- The invention of the printing press brought information to many, and the telegraph system got information to others faster across vast distances, through to its use in the UK coding-breaking machines such as Colossus and the German encoding machines such as Enigma;
- And, finally, to the twenty-first century with its cyber warfare, Big Data technology, Internet of Things (IOT), cloud, smart cities/infrastructure, social media and personally identifiable information (PII).

The list of examples on the use and importance of information is as endless as history itself. Today, we say we live in a global "information society" that is constantly expanding and shaping the way we think, work, act and play.

Information from ancient times to now (and for the future) has and will continue to be a valuable item to possess, and so it has always been and will continue to be the target of risk:

- Risks to a person's information (e.g., disclosure and theft of information about his personal estate and affairs, his health, his discoveries and inventions, modifications of his plans, loss or corruption of his personal data that protects image and reputation),
- Risks from information used by his opponents (e.g., a competitor knows more than he does, traders in the Renaissance times knew a better trading route without getting lost or falling foul of rocks and uncharted waters,[2] Sun Tzu's Art of War and business strategy);
- Risks from sharing information with others (e.g. a partnership—business contracts, friendship—or social media).

The very same market innovations, products and services that have shaped this digital world we are a part of and the benefits we have gained

2. The Spanish and Portuguese explorers and seafarers used notion of risk in the sixteenth and seventeenth centuries to refer to sailing into uncharted waters. This use of risk is associated with a notion of uncertainty related to space. We then adopted a notion of risk related to time, as used in financial processing and investment. In this later case, it is related to the probable consequences of investment decisions for borrowers and lenders. Today the notion of risk relates to a wide range of other space/time situations of uncertainty, including risks to our organisations' financial and nonfinancial assets.

from this digital world have also given us a world where the information we use, value and rely on is at risk and is becoming more and more difficult to protect. The importance and value of information and protecting this value is the central theme of this book.

1.2 Globally Connected

> *Thought is an idea in transit, which when once released, never can be lured back, nor the spoken word recalled. Nor ever the overt act be erased.*
>
> —Accredited to Pythagoras

Today's so-called information society, digital economy, digital agenda or whatever other name we might choose to describe the phenomena that is dominating our world is increasingly driving, shaping and reforming the way we work, live and play. Organisations around the world are doing business across public and private networks. Every type of business, from the very small to the very big, is affected by the information society. Citizens in many countries of the world are connecting to the Internet doing online shopping, banking and hotel and airline reservations. Industry and governments alike are seeing the benefits of going down an electronic commerce route and reaping the benefits of the digital economy.

Today's technologies make it easier to have access to information, to store more of it, to process it faster and to communicate it globally in seconds. Mobile and smart technologies make it easier to stay globally connected all the time, and the exponential growth in mobile apps is transforming our lives like never before, enabling us to do more on the move. This capability gives us the means of being more informed and more capable of exploiting and taking advantage of a vast range of business and private opportunities. But the same technology can at the same time be used to damage or destroy information.

With the emergence of the IoT, users via not just laptops, mobile and smart devices, but many billions of small sensors and other devices connected through wireless networks have also connected to the Internet, making the size of the Internet community bigger. These things are, for example, helping to control smart energy, transport and building systems and so are collectively involved in sensing, communicating and in a limited way processing information—which may be critical, sensitive or even personal.

The diverse range of network technologies and services available today make it easy to access the world, to be globally connected and to be online at any time, from almost anywhere. This makes information more vulnerable, whether it is business, government, personnel or private information.

The email to a friend or business colleague once sent is public—it can be deleted by the receiver, but it cannot be recalled. Remarks on a social media site can become viral within a short period of time. The power of the Internet and global interconnectivity has given many benefits to our work and our life styles—it has also made many of things we do more public and less private, and it has raised the need for more awareness, care and attention to how we should use this global connectivity and how we should behave in a responsible, safe and secure way.

1.3 More Ado About Risks

As a rule, men worry more about what they can't see than about what they can.
—Accredited to Julius Caesar

Risk is about the effect of uncertainty on objectives, so says ISO 31000. What is the uncertainty of whether your organisation is going to be infected by a malware attack? If it happens what is the effect on the organisation's security objectives—maybe information will be stolen by competitors, which will have a financial impact on the organisation, or maybe information is modified or destroyed and the impact is loss of productivity whilst the information is restored.

Organisations' dependency, use and application of information are all pervasive—all IT-based business processes are driven by the supply of information, and the output of this process is yet again information. This dependency on information is made more urgent by the potential risks to this information and the uncertainty of the risk. Understanding what the information security risks are and assessing how these risks have an effect and impact business are vital to being able to manage these risks effectively.

An organisation needs to assess its risks in terms of the impacts resulting from a security incident happening. It needs to adopt an approach to risk assessment that is suitable to its business environment, and to invest in the management of risk and to minimize any business impact. The risk management process needs to minimize the risks (*effects of downside risks*) by making well-informed decisions of how to treat the risks in the most cost-effective manner whilst taking account of the need to support and maximize the business opportunities (*opportunities through upside risks*). The risk management process involves taking the necessary action to ensure that the options to treat the risks are implemented whilst supporting the business opportunities.

It is important that an organisation deals with information security at all levels to ensure availability and continuity of business resources, to treat the risks to avoid or reduce any potential damage and impact to the business.

1.4 Decoding the Secret of Information Security Management

The secret of change is to focus all your energy, not on fighting the old but on building the new.

—Accredited to Socrates

Management systems like the information security management system (ISMS) involve cycles of continued improvement to deal with changes and to maintain a level of effectiveness suitable and appropriate to the organisation and its business. The ISMS ethos is about keeping up to date to make improvements and treat the risks associated with changes.

Uncertainty and change are inevitable, inescapable and universal aspects of life and the world we live in. Whether in business or personal life, there are uncertainties and there are changes—"nothing is as permanent as change."[3] The definition of risk as discussed in Chapter 2 centres around the notion of uncertainty, and all changes (expected or unexpected) carry an element of risk. Information security risks threaten the digital world we live in and as this digital world evolves and changes the risks change and evolve. As businesses and consumers engage in this digital world, they need to confront the uncertainties of the risks that are involved. The "secret" of information security management systems lies in the continual improvement processes involved, which enable businesses to manage the changes and risks and keep up to date with their risk treatment programme—applying effective security process and controls to protect their information. This means preserving the confidentiality, integrity and availability of information by managing the risks. These three security objectives should be the focal point of the organisation's information security policy. In addition, the organisation needs to take care of the personal data, or PII (personally identifiable information), it processes relating to its staff, customers and consumers. The ISMS that the organisation implements needs to continually evolve and improve with the pace of change to achieve these objectives.

The digital revolution as a way of doing business has been growing by leaps and bounds, and it has become the standard way of doing business. Technologies such as the cloud, Big Data and IoT are making our world more connected, smart and risky. Standards, such as the ISO/IEC 27001 family, provide a means of managing the information security risks in our cyber world. This family of standards if properly implemented can give support to

3. Accredited to Heraclitus: "All is flux nothing is fixed."

addressing questions such as "Is my organisation (are my trading partners) fit to do online business in a secure way?" or, for assurance associated with publicly available systems, "Is my organisation's online site secure?"

Other questions involve confidence in third party arrangements: "Are they fit to have access to my organisation's information processing facilities and to share information with?" or "Are they fit to look after my organisation's information and information processing facilities?"

1.5 Management and Awareness

Education is kindling the flame not filling the vessel.

—Accredited to Socrates

The emphasis of ISO/IEC 27001 is the protection of information and what underpins the success of having an effective, suitable and appropriate ISMS implementation is management. This entails the scope of management in its broadest sense—from the CEO and others in the C-suite category through to other layers of management, it involves the management of people, policy, procedures, processes and systems of risk management controls and information, and to be a comprehensive framework for managing information security it also includes the management of everything that supports the business—the services, business applications and IT.

Management is the heart and soul of a successful ISMS implementation. Direction needs to driven by top management policy, there needs to be active commitment and leadership for the ISMS and this needs to come from the top. Resources need to be made available and committed to from the top of the organisation. Management needs to engender and promote an information security culture in the organisation to support its use of its ISMS.

Awareness and education is critical to having an ISMS that will be effective in managing the risks. Using the organisation's ISMS in practice, day by day, week by week, as part of managers' and employees' work function is important. Awareness and education should, in this sense, be about the practice of using the ISMS, not about the theory of the ISMS. For managers and employees, it should be about using and applying the ISMS policies and procedures to their particular work circumstances. What should they do, and how should they respond to security events as they happen? What is the proper way to secure information they handle? Do they know which procedure to use for a particular work task that involves some security aspect? These and many other day-to-day practical aspects of carrying out their specific job function in a manner that will protect the information they are dealing with is at the heart of good information security awareness. So it is not just about building up one's knowledge, but understanding how to put this knowledge into practice, knowing how to do things in practice and knowing how to react and respond to information security risks. This practical understanding might make the difference between the organisation successfully surviving a malicious attack or suffering untold damage because of this attack.

Information security is teamwork, and everybody plays an important role in managing the risks—from the CEO down to employees should all be aware of their responsibilities and what they need to do and to contribute toward achieving an effective ISMS.

1.6 Legislation, Regulation and Governance

> *Good people don't need laws to tell them how to act responsibly . . .and bad people will find a way round the laws.*
>
> —Accredited to Plato

Over the past three decades or so, more and more attention has been given to legislation, and regulation in various areas are related to some aspect of information (e.g., data protection and privacy legislation, computer misuse and hacking, social media, electronic commerce and the use of electronic signatures, digital evidence or cyber security). It is very likely that the future will see more and more such regulations appear.

Laws and regulations have a business impact on any organisation—they clearly need to comply with such legislation and demonstrate their compliance to such legislation.

The growth in online business has further increased the need for legal protection to protect businesses and their transactions as well as PII that is stored and processed by servers in different jurisdictions.

Some examples of current legislation include the following:

- SoX or Sarbox (Sarbanes Oxley) also known as the Public Company Accounting Reform and Investor Protection Act of 2002 (US)
- Cyber security–related regulations in the US include the 1986 Computer Fraud and Abuse Act, the 1996 Health Insurance Portability and Accountability Act (HIPAA), the 1999 Gramm-Leach-Bliley Act (GLBA), the 2002 Homeland Security Act and the 2013 Cyber Intelligence Sharing and Protection Act (CISPA); the UK Computer Misuse Act criminalises unauthorised access to computer systems;
- The EU Directive on Data Protection (plus the legislation of each Member State—their national interpretation of the EU Directive) cov-

ers the gathering, handling and exchange of personal data; there is data protection/data privacy legislation in non-EU countries including Australia, Canada, Hong Kong, Malaysia, Mexico, New Zealand, Norway, Republic of South Korea, Singapore, South Africa, Switzerland;

- The EU Directive on Electronic Signatures (plus the Member States interpretation) provides legal provisions regarding the legal admissibility of electronic signatures (e.g., those used for online transactions) and other similar pieces of legislation appear in other parts of the world such as in Hong Kong and Japan;
- International copyright and licensing laws provide legal measures against piracy and theft of intellectual property.

1.7 En Route to a Certified Business Environment

We acquire a particular quality by acting in a particular way.

—Accredited to Aristotle

The quality of our information security depends the actions we take to achieve an effective ISMS. In more than 100 countries around the world, organisations are using ISO/IEC 27001 to manage their information security by implementing an ISMS. In addition, many of these organisations have opted to have their ISMS certified by a third- party certification body. This certification process against ISO/IEC 27001 provides a common international benchmark for assessing an organisation's information security management system. Certification reassures management, customers, suppliers and business partners that an organisation is applying best information security practice. It opens up new avenues of business based on a protected business environment.

ISO/IEC 27001 and the process of certification (see the accredited certification standard ISO/IEC 27006 for more details) provides a tried and tested framework, a metric system for determining whether your business is "fit for purpose" from an information security management perspective. The certification audit will check that the ISMS if in conformance with the requirements in ISO/IEC 27001. The audit team will ask many questions and seek evidence to check conformance.

1.7.1 Processes

- In the planning, designing and development of the ISMS have you taken account the requirements and expectations of all interested parties and any issues (internal or external) that might be relevant?
- Do you have an effective and systematic risk assessment process in place? Have you implemented an appropriate set of controls to treat the identified, assessed, and evaluated risks?
- Do you have effective, adequate, and suitable management processes in place to govern and manage the ISMS?
- What processes do you have in place to check, assess and verify the performance of your implemented ISMS?
- Do you have a continual improvement process in place? Does this adequately manage and effectively adapt to changing circumstances that are relevant to the ISMS?

1.7.2 Controls

- Do you have an information security policy? Do you have all the necessary procedures in place? Are they sufficient?
- Are your employees aware of and trained in security for the organisation in general and their job in particular?
- What controls and measures are in place to handle information securely? How are you controlling access to your sensitive and critical commercial information? Do you protect the integrity and availability of your business information and processes?
- Have you the controls and procedures in place to securely use, manage and monitor your operations, networks and communications?
- Are you protected from physical and environmental risks?
- Do you have business information security continuity plans in place?
- Are you compliant with all the laws and regulations relevant to your business?

Establishing a certified ISO/IEC 27001 business environment to demonstrate confidence to the management board, C-suite managers, your business partners and customers that your organisation is "fit to do business with" is common among many thousands of organisations around the world—small-, medium- and large-sized organisations in most business and industry sectors as well as governments and academic and research institutions.

CHAPTER

2

Contents

ISO/IEC 27001 ISMS Family

2.1 ISO/IEC Standardisation

2.1.1 Overview

The work in International Standards Organisation (ISO) and International Electrical Committee (IEC) is carried out by Technical Committees (TCs) and Subcommittees (SCs). These committees are responsible for the executive decision-making and overall management of the standards programme. In addition Working Groups are established in the SCs to carry out the development of the standards. The Joint Technical Committee (JTC 1) is a joint ISO and IEC committee responsible for IT-related standards. For more detailed information about the standardisation process; who gets involved in the work; the technical committees and what they do and the deliverables (standards, technical report), the reader is directed to the ISO web site at www.iso.org/iso/home/standards_development.htm.

This chapter provides an overview of the work in ISO/IEC JTC 1 related to information and IT security standardization, and in particular information security management system standards and the subcommittee SC 27.

2.1.2 ISO/IEC JTC 1/SC 27

2.1.2.1 SC27 Structure

ISO/IEC JTC 1/SC 27 is the committee responsible for a wide range of information and IT security standards projects. The subcommittee consists five working groups (WG 1–WG 5) covering a diverse ranges of information and IT security subjects. Figure 2.1 shows the current structure of SC27. The SC 27 working group WG 1 is particularly responsible for the ISMS ISO/IEC 27001 family of standards.

2.1.2.2 SC27 Membership

The membership of SC 27 currently includes national standard bodies representing 50 different countries. These are voting members in addition to national standard bodies representing 20 different countries, which are are observing members (nonvoting).

SC 27 also works in cooperation with many other organisations, normally referred to liaison bodies, in order to facilitate the exchange of information on future needs for security standards as well as engaging in collaborative work on the development of the committee's standardization projects. The liaison body may be another committee within ISO, IEC or JTC 1, or an organisation outside of the ISO and IEC committee structure. This might be another international standards organisation, such as ITU-T, or a regional standards body such as the European Telecommunications Standards Institute (ETSI), or it may be an organisation representing an industry sector, consumer group or user association. Currently (as of September 2015), SC 27 has 71 national body members (these are the national standards organisations from 71 different countries) and more than 55 liaison members (these are liaison organisations that also contribute to the work). The large number of members of SC 27 means a wealth of contributions, knowledge, expertise and skills of professionals in the field of information security is available from around the globe. This also means the needs and expectations of organisations and businesses working in most market sectors contribute to the development of the final product. This wealth of input, as well as the review and feedback of contributions, adds to the value of the standards developed.

2.2 Overview

2.2.1 International Standards

The ISO/IEC 27001 family of information security management system (ISMS) standards is developed at the international level in response to market needs. Like all ISO and IEC, they are based on global expert opinion and

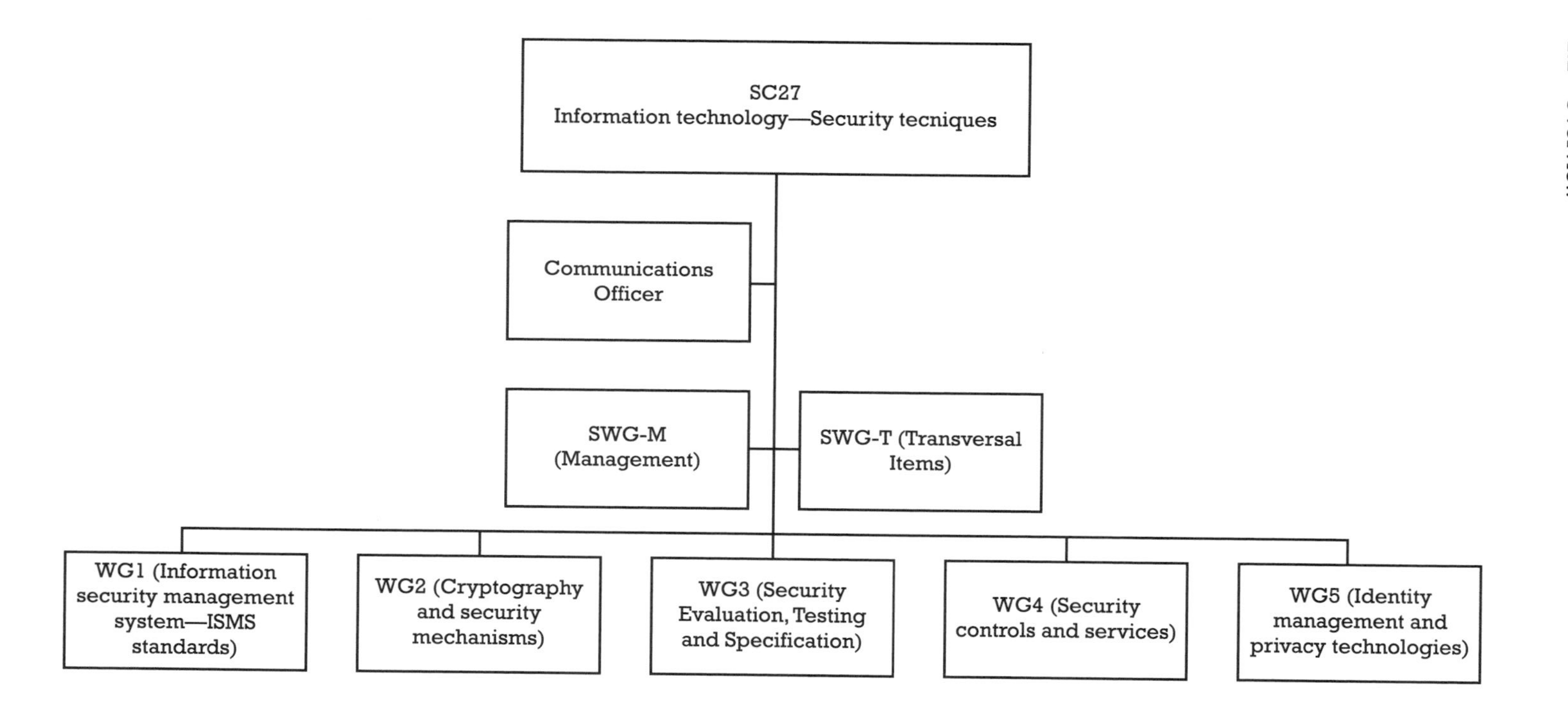

Figure 2.1 Structure of SC27.

developed through a multistakeholder process, using a consensus-based approach. All such standards go through a maintenance life cycle to keep them current and up to date—this means that standards are often revised after a period of time, normally about five years. Therefore, a standard may have many editions, as each new edition replaces the previous edition. Users of ISO/IEC standards should always check if the version they are using is the most current edition. Users of ISO/IEC standards should, therefore, always check if the version they are using is the most current edition, as revised versions are likely to be developed over time.

This book addresses the requirements of the second edition of ISO/IEC 27001. All the other members of the ISO/IEC 27001 family that are quoted are the current versions at the time of writing this book. These versions may be replaced with new editions in the fullness of time, including the second edition of ISO/IEC 27001.

2.2.2 The 27001 ISMS Family

This chapter presents details of the current series of information security management system (ISMS) standards called the ISO/IEC 27001 family:

2.2.2.1 ISMS Requirements

ISO/IEC 27001: Information security management system requirements: This is the core standard in the ISO/IEC 27001 family. This standard specifies the requirements for establishing, implementing, deploying, monitoring, reviewing, maintaining, updating and improving information security management systems (ISMS).

2.2.2.2 ISMS Supporting Guidelines and Code of Practice

- ISO/IEC 27002: Code of practice for information security management;
- ISO/IEC 27003: ISMS implementation guidelines;
- ISO/IEC 27004: Information security management measurements;
- ISO/IEC 27005: Information security risk management.

This group of standards in the ISMS family provides advice, guidance and implementation to support for the purposes of meeting the requirements in ISO/IEC 27001.

2.2.2.3 ISMS Accredited Certification and Auditing Standards

- ISO/IEC 27006: International accreditation guidelines for the accreditation of bodies operating certification/registration of information security management systems;
- ISO/IEC 27007: Guidelines for information security management systems auditing;
- ISO/IEC 27008: Guidelines for auditors on ISMS controls;
- ISO/IEC 27009: Sector-specific application of ISO/IEC 27001—requirements;
- ISO/IEC 27021: Competence requirements information security management professionals.

This group of standards covers requirements for third-party accredited ISMS certification audits (ISO/IEC 27006), audit guidelines covering first, second and third party audits (ISO/IEC 27007) and a technical report giving guidance for assessing the implementation of ISMS controls selected through a risk-based approach for information security management (ISO/IEC 27008). The group also includes a standard (ISO/IEC 27009) that focuses on sector specific application of ISO/IEC 27001. Finally, a new departure is a standard (ISO/IEC 27021) that deals with competence requirements to support certification schemes dealing with the certification of information security management professionals.

2.2.2.4 ISMS Sector Specific

- ISO/IEC 27010: Information security management for intersector and interorganisational communications;
- ISO/IEC 27011: Information security management guidelines for telecommunications organisations based on ISO/IEC 27002;
- ISO/IEC 27013: Guidelines on the integrated implementation of ISO/IEC 27001 and ISO/IEC 20000-1;
- ISO/IEC 27015: Information security management guidelines for financial services;
- ISO/IEC 27017: Guidelines on information security controls for the use of cloud computing services based on ISO/IEC 27002;

- ISO/IEC 27018: Code of practice for PII protection in public clouds acting as PII processors;
- ISO/IEC 27019: Information security management guidelines based on ISO/IEC 27002 for process control systems specific to the energy utility industry.

This group of standards covers sector-specific application of ISO/IEC 27001 and ISO/IEC 27002. This includes defining sector-specific controls and requirements in addition to those defined in ISO/IEC 27001 and ISO/IEC 27002, as well as providing additional implementation guidance for sectors where necessary.

2.2.2.5 ISMS Family Support

- ISO/IEC 27000: ISMS overview and vocabulary;
- ISO/IEC 27023: Mapping revised editions of ISO/IEC 27001 and ISO/IEC 27002;
- ISO/IEC 27014: Governance of information security;
- ISO/IEC 27016: Information security management—organisational economics.

This final group covers a miscellaneous set of subjects: a standard that covers the main items of terminology used in the ISMS family together with an overview of the family. This standard considers the governance aspects of information security (ISO/IEC 27014) through which organisations direct and control the information security management system (ISMS) process as specified in ISO/IEC 27001 and a technical report that covers the important, but often forgotten, topic of economics of information security (ISO/IEC 27016). This provides guidelines on how an organisation can make decisions to protect information and understand the economic consequences of these decisions in the context of competing requirements for resources. ISO/IEC 27023 provides a set of maps, which show the transition from the 2005 to the 2013 editions of ISO/IEC 27001 and ISO/IEC 27002.

2.2.3 Standards Interrelated to 27001 ISMS Family

A number of additional standards are not included in the family listed in 3.1.2. They are nevertheless related to the ISO/IEC 27001 family. Many of these standards provide more detailed implementation information. The current set of standards are summarised as follows:

2.2.3.1 ICT Readiness, Information Security Incident Management, IDPS, Digital Investigations

- ISO/IEC 27031: Guidelines for ICT readiness for business continuity;
- ISO/IEC 27035: Information security incident management;
- ISO/IEC 27037: Guidelines for the identification, collection, acquisition and preservation of digital evidence;
- ISO/IEC 27039: Selection, deployment and operation of intrusion detection and prevention systems (IDPS);
- ISO/IEC 27041: Guidance on assuring suitability and adequacy of incident investigative methods;
- ISO/IEC 27042: Guidelines for the analysis and interpretation of digital evidence;
- ISO/IEC 27043: Incident investigation principles and processes;
- ISO/IEC 27050: Electronic discovery.

2.2.3.2 Applications and Services

- ISO/IEC 27032: Guidelines for cybersecurity;
- ISO/IEC 27033: Network security;
- ISO/IEC 27034: Application security;
- ISO/IEC 27036: Information security for supplier relationships;
- ISO/IEC 27038: Specification for digital redaction;
- ISO/IEC 27040: Storage security.

In addition to the standards mentioned in Section 2.2.2 and 2.2.3, SC 27 develops many other standards (e.g., in the field of cryptographic techniques, standards for security evaluation criteria for products and systems of products and standards for identity management and privacy management). Figure 2.2 shows a high- level view of the range of standards developed in SC 27 arranged to reflect how they relate to ISO/IEC 27001.

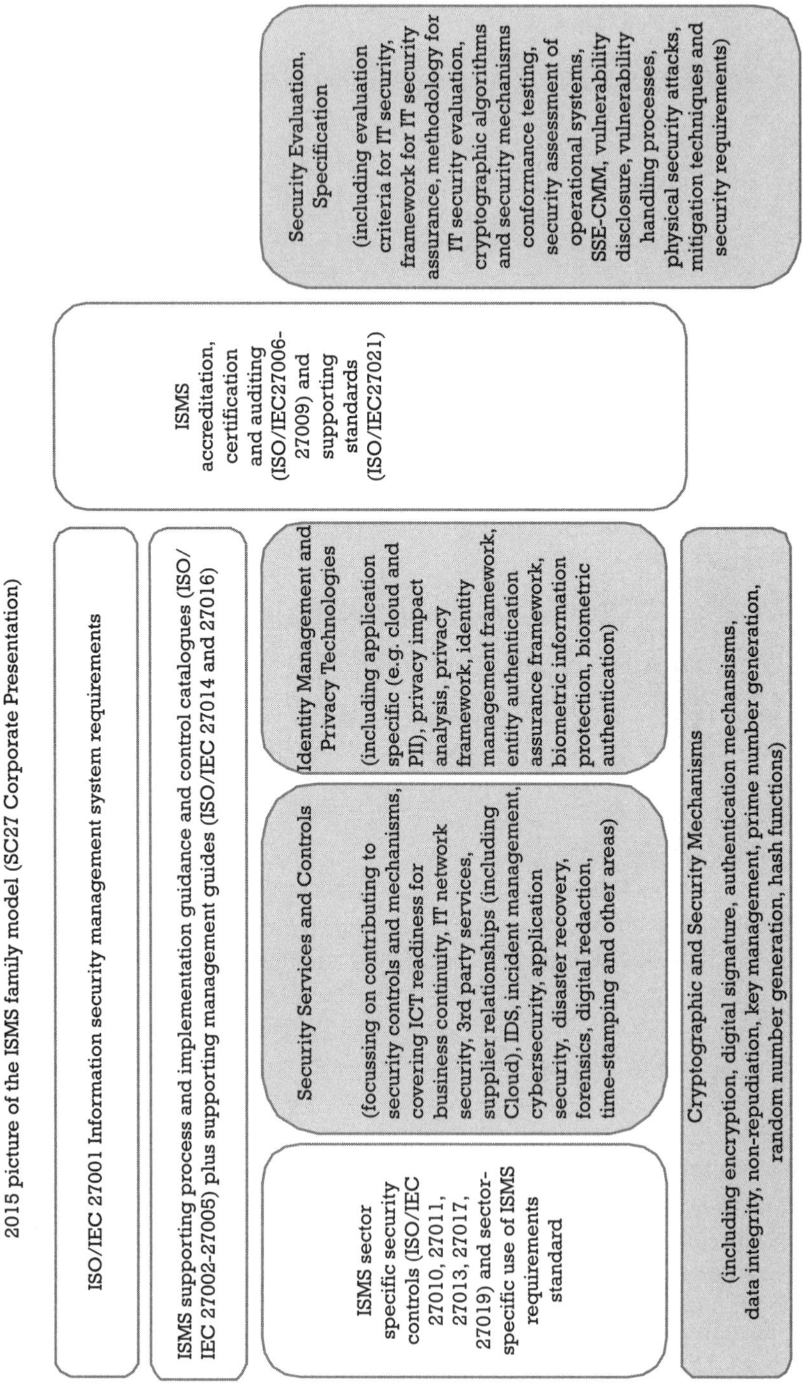

2015 picture of the ISMS family model (SC27 Corporate Presentation)

ISO/IEC 27001 Information security management system requirements

ISMS supporting process and implementation guidance and control catalogues (ISO/IEC 27002-27005) plus supporting management guides (ISO/IEC 27014 and 27016)

ISMS sector specific security controls (ISO/IEC 27010, 27011, 27013, 27017, 27019) and sector-specific use of ISMS requirements standard

Security Services and Controls

(focussing on contributing to security controls and mechanisms, covering ICT readiness for business continuity, IT network security, 3rd party services, supplier relationships (including Cloud), IDS, incident management, cybersecurity, application security, disaster recovery, forensics, digital redaction, time-stamping and other areas)

Identity Management and Privacy Technologies

(including application specific (e.g. cloud and PII), privacy impact analysis, privacy framework, identity management framework, entity authentication assurance framework, biometric information protection, biometric authentication)

Cryptographic and Security Mechanisms

(including encryption, digital signature, authentication mechansisms, data integrity, non-repudiation, key management, prime number generation, random number generation, hash functions)

ISMS accreditation, certification and auditing (ISO/IEC27006-27009) and supporting standards (ISO/IEC27021)

Security Evaluation, Specification

(including evaluation criteria for IT security, framework for IT security assurance, methodology for IT security evaluation, cryptographic algorithms and security mechanisms conformance testing, security assessment of operational systems, SSE-CMM, vulnerability disclosure, vulnerability handling processes, physical security attacks, mitigation techniques and security requirements)

Figure 2.2 Standards architecture from an ISMS perspective.

2.3 Evolution of the ISO/IEC 27000 Family

2.3.1 The Weakest Link

Managing information security from a people, policy, procedural and business process point of view is the core objective behind approach taken in the ISO/IEC 27000 family of ISMS standards. Although we are highly dependent on technology and the role it plays in today's business environment as our information-processing workhorse, this is just one aspect and IT security is not the most significant challenge to be faced. People themselves are probably present the major problem to information security. People are the greatest of all vulnerabilities, and the management of people and the processing of information presents the greatest of challenges. People are risk takers when it comes to a range of everyday tasks they are involved in, and at the same time they are risk averse in other tasks they perform. Training, investing and motivating people, as well as allocating responsibilities for security and making them feel that they are part of a security culture, are some of the key parts of establishing an effective awareness and management system that will help businesses protect their information. Recognition of the significance that people, procedures and processes are the greatest problems to resolve in the domain of information started to grow in the late 1980s but did not become a serious topic of standardization until the mid-1990s, and certainly it was not until the 2005 onwards that it started to become an universal agenda item on management review and board-level meetings.

2.3.2 Baseline Controls

Best practice controls have been the essential element of information security management and the prehistory of the ISO/IEC 27000 of ISMS standards, going back as far as the late 1980s when it really started to become of age. More and more use of controls commonly used by business was brought together: controls that business could employ without the need to undertake any costly commitment. Businesses could set a level of protection across their organisation using these "common use" best practice controls to establish a "baseline" level of security as a common security standard. In addition, of course, businesses had to build upon this baseline where security controls dealing with specific risks were necessary.

Catalogues of baseline controls were produced in many business sectors and user groups such as the International Information Integrity Institute (I4). The I4 work adopted those controls in common use by industry as per the criterion "if the majority of organisations uses a specific security control, then it is defined as control in 'common use'" and is thus a baseline control.

In the early 1990s, the UK, in the Department of Trade and Industry (DTI), set up an industry group to establish a code of practice: a code developed by industry, for industry. This code was a catalogue of best practice security controls, including some of the baseline controls discussed and adopted by industry in the 1980s. However, this code went one step further introducing the notion of risk assessment as a way of matching the controls to the business value of the information and information system of each organisation. This code would thus allow organisations to customize their information security to the needs of their business: specific information systems they are using, their business processes and applications, and their trading and operational environment. This DTI code also included advice and guidance on the implementation of the controls. This code was published in 1992 by the DTI.

The idea of applying a risk assessment was also introduced into the code as a management tool for establishing the security requirements of the business and for selecting a set of controls to match these requirements.

2.3.3 Formative Years—BS 7799 Part 1 and Part 2

In 1995 the DTI code of practice was published as a British Standard BS 7799: 1995. In 1998 it was decide to carry out a review of the 1995 version to check whether there was a need to revise it or to leave it as is: this is normal practice with all standards. This led to a decision to revise the standard to improve and update it whilst retaining backward compatibility with the 1995 version. An editing team under the management of the BSI committee BDD3 set about reviewing and collecting comments from a range of interested parties.

From 1998 onwards, a family of BS 7799 standards was then progressed:

- BS 7799 Part 1 of this family was the DTI code of practice for information security management;
- BS 7799 Part 2 is a specification for an ISMS. This development arose after a public consultation on the need for a third party certification scheme for ISMS. The certification and audit process model used for BS 7799 Part 2 is the same as that used for ISO 9001 for quality and ISO 14000 for environmental management systems.

The emergence of the second part of the standard BS 7799 marked a second significant milestone in information security management. The idea of applying a risk assessment, although mentioned in Part 1, is specifically introduced in Part 2 as a mandatory requirement as a means of establishing the security requirements of the business and for selecting a set of controls to match these requirements.

Organisations could establish and implement an ISMS based on Part 2 and subsequently went for third party certification of their ISMS to demonstrate that their information security was "fit for purpose." Such certification provides a means for measuring the effectiveness of the ISMS giving assurance to their customers, business investors, shareholders and trading partners.

The family of BS 7799 provides best practice suitable for organisations, whether small, medium or large and in every market sector, including government and commercial organisations. They give organisations best practice for use in various trading relationships including customer-supplier chains, collaborative ventures, third-party services, outsourcing arrangements and virtual private networks of organisations distributed around the world, as well as the emergence of global B2B electronic commerce and as G2B, G2C and B2C relationships.

2.3.4 Internationalization

Up until 2000 these standards were being used worldwide by many different industries and businesses, and they became de facto international standards from an industry perspective. The next stage in the development of these standards was to formalize them as international standards. This led in 2000 to the proposed introduction of BS 7799 Part 1 into ISO/IEC JTC 1.[1] The proposed introduction into ISO/IEC achieved the minimum majority support (67%) to be approved. Since not all member countries of JTC 1 gave a vote of approval, and it is always good to achieve a consensus, it was decided after some debate that Part 1 would be published under the condition that an early revision of the standard should commence as soon as possible. So in 2000 BS 7799 Part 1 became ISO/IEC 17799:2000.

In 2005 a revised version of ISO/IEC 17799:2000 was published and in the same year BS 7799 Part 2 became ISO/IEC 27001: 2005. Also in the same year, SC 27/WG 1 adopted the ISO/IEC 27000 numbering scheme, and the ISO/IEC 27000 family of ISMS standards was adopted.

Today, the international community is now adopting the ISO/IEC 27000 family as the common language for information security. This enables organisations worldwide to engage in securing their business using such a language to demonstrate to their customers and business partners that they are fit for purpose to handle information in a secure way, whether it be online business or offline business. The world is now opening up to this notion of a

1. Joint Technical Committee 1 (JTC 1) is responsible for IT-related standards.

common ISMS language for the benefit of all organisations to manage their risks and to protect one of their critical assets—information.

2.4 Overview of ISO/IEC 27001: 2013

2.4.1 Introduction

The international standard ISO/IEC 27001 is an ISMS set of requirements for establishing, implementing, deploying, monitoring, reviewing, maintaining, updating and improving a documented ISMS with respect to an organisation's overall business risks and opportunities.

It belongs to a class of standards referred to as the Management System Standards (MSS), which includes standards such as ISO 9001(Quality Management System), ISO 14001 (Environmental Management System), ISO 22000 (Food Safety Management System), ISO/IEC 20000-1 (Service Management System) and ISO 22301 (Business Continuity Management System).

In 2012, ISO published a common approach (ISO Directions Annex SL, Appendix 3) for both the development of new MSS and for the revision of existing MSS. The reasons for this were to enable an organisation to operate an integrated MSS that will comply with the requirements of two or more MSS. For example, an organisation could decide to operate with both ISO/IEC 27001 and ISO 22301, which would mean they would integrate their ISMS with the BCMS. Using this common integrated approach would provide many business benefits and increase the value of using MSS. Both ISO/IEC 27001 and ISO 22301 are two standards from the MSS family of standards that have been revised to take on this common approach, which involves using a high-level structure, identical core text and common terms and core definitions.

The second edition of ISO/IEC 27001 was published in 2013 following a three-and-a-half year revision cycle. This new version takes account the new MSS approach. This means that the high-level structure of the chapters, clauses and sections looks different than the 2005 edition. In addition to changing the high-level structure, changes were made to the requirements specified in the standard. These changes reflected the contributions received from member bodies of SC 27 and their cooperating organisations. ISO/IEC 27023 is a guide that provides transition maps showing the high-level changes that have been made between the 2005 and the 2013 editions of both ISO/IEC 27001 and ISO/IEC 27002. This guide is very useful for those wanting to know in more detail where the changes have occurred.

This section of Chapter 2 provides some of the highlights of the second edition of ISO/IEC 27001. More specific discussion on the second edition can be found in later chapters. These subsequent chapters will cover the changes in more detail.

2.4.2 ISMS Audience

The organisational target audience of ISO/IEC 27001 has not changed in the 2013 edition: it is suitable to all types and sizes of organisations. It can be applied to any type of business activity and across all business markets, since its subject matter is the protection of information, irrespective of what systems, processes or IT the organisation deploys.

The functional target audience in the 2013 edition has also not changed. The second edition of ISO/IEC 27001 places more emphasis on the role of management, leadership and commitment of management in supporting the ISMS. There is more emphasis on the need to align the development of the ISMS with the needs and expectations of stakeholders and all relevant interested parties, and to make sure all internal and external business issues and requirements are covered.

2.4.3 Mandatory Statements

The second edition of ISO/IEC 27001 still uses the word "shall" in specifying the requirements, and in ISO terminology any requirement that includes this word is mandatory to implement if an organisation wishes to claim conformance with the standard. Therefore, this means that this standard can be used for formal third-party certification, which is similar to the ISO 9001 case for quality management systems.

2.4.4 Processes

The ISO/IEC 27001: 2005 was based on a Plan-Do-Check-Act process model. In the 2013 edition of ISO/IEC 27001 this model has been excluded, although its continual improvement philosophy is certainly still firmly in place. The process-based approach, however, is still very much a part of the new edition of ISO/IEC 27001, as was the case with the old edition. For example, the organisation needs to have a risk assessment process to be implemented or risk assessment process or an internal audit process. ISMS processes are the systematic operations and activities that are a central feature of ISO/IEC 27001.

2.4.5 ISMS Stages

The ISMS stages are establishing, implementing, deploying, monitoring, reviewing, maintaining, updating and improving and the organisation needs to go through a number of staged activities. These stages include a number of shall requirements (mandatory requirements) where things need to be done, activities need to be carried out and processes need to be implemented. These requirements fall under the following clause headings:

- The context of the organisation (Clause 4);
- Leadership (Clause 5);
- Planning (Clause 6);
- Support (Clause 7);
- Operation (Clause 8);
- Performance evaluation (Clause 9);
- Improvement (Clause 10).

2.4.6 Risk-Based Approach

The purpose of the risk-based approach is to take care of the information security aspects of the organisation's business activities. The ISMS risk management process needs to take into account the requirements and expectations of all interested parties, including customers, consumers and business partners. It needs to take into account any issues that might be relevant to information security risks, be they related to corporate governance, legal, regulatory and contractual obligations, business objectives and strategy, business operations and processes or the use and application of information and communications technology (ICT) systems.

The overall risk philosophy in the new addition is based on the concepts and terminology defined in the generic risk standard ISO 31000. The clauses in this chapter are a mere overview of the more detailed discussion on ISMS risk management found in Chapter 4.

Both in ISO/IEC 27001: 2005 and ISO/IEC 27001: 2013, risk management is a central theme; however, the 2013 edition includes a number of important changes to the risk management process.

2.4.6.1 Risk Assessment

One of the significant changes in requirements between the 2005 and 2013 editions is the need to identify the assets, threats and vulnerabilities. This is no longer a requirement in the second edition. So the asset-based approach in 2005 has been replaced with an approach based on the model defined in ISO 31000. This current risk assessment will be discussed in Chapter 4.

2.4.6.2 Risk Treatment

Another change to be found in the second edition is related to the treatment of risk. In the 2005 edition Annex A was used to select an appropriate set of controls from to reduce identified risks. In the 2013 edition the user determines a set of controls in accordance with the risk treatment options that the organisation has decided to implement. The organisation then needs to compare this set of controls with the Annex A controls to benchmark whether any important controls have been excluded.

The standard ISO/IEC 27005 provides guidance on the information security risk management in support of ISO/IEC 27001.

2.4.7 Performance Evaluation

In the 2005 version of ISO/IEC 27001, performance evaluation was considered and implemented through the use of several processes including taking measurements, monitoring, internal audits and management reviews. In the 2013 edition these same processes are specified and invoked; however, they have been brought together in a single chapter and the wording of the content has undergone some improvements. The standard ISO/IEC 27004 provides guidance on the requirements information security measurements given in ISO/IEC 27001.

2.5 Second Edition of ISO/IEC 27002

At the same time that ISO/IEC 27001 was being revised, so was the standard ISO/IEC 27002 code of practice for information security management being revised. The revised versions of these standards were released at the same time.

The changes to ISO/IEC 27002 included the deletion of some controls, the addition of some new controls and the modification of controls from the prevision edition. ISO/IEC 27023 is a guide that provides transition maps showing the high-level changes that have been made between the 2005 and the 2013 editions of ISO/IEC 27002.

2.5.1 Conformance with ISO/IEC 27002

The term "conformance" is often misunderstood and sometimes confused and used interchangeably with the word "compliance." The code of practice ISO/IEC 27002 takes the form of guidance and recommendations, as such, it is not a conformance assessment standard, using the ISO technical use of the term in the sense of a management system standard, as it uses "should" statements, unlike ISO/IEC 27001, which uses "shall" statements. Care needs to be taken to ensure that claims of conformance are not misleading.

2.5.2 Applying ISO/IEC 27002

ISO/IEC 27002 is primarily a catalogue of best practice controls, which users can select from to deploy security management controls in their business environment to achieve a baseline of best practice protection. When combined with ISO/IEC 27001, these two complement each other, providing organisations with a set of tools for managing information security risks (see Chapter 4 for the change in how Annex A of ISO/IEC 27001 is now used).

Of course, ISO/IEC 27002 can be used on its own, but this is outside the management system risk-based processes specified in ISO/IEC 27001, which are there to facilitate the management of an effective information security system with a built-in programme for continually improving an organisation's security status. Subsequent chapters in the book will provide more detailed information on the use, implementation and application of the ISO/IEC 27002.

CHAPTER

3

Contents

ISMS Business Context

3.1 Organisational Context

3.1.1 Understanding the Business

Before embarking on the development and implementation of an ISMS, it is important to understand your organisation, its context and purpose, its organisational structure and how it works and operates. This includes consideration of the issues, both internal and external, that are relevant to the organisation's purpose and business objectives and that could influence the ISMS outcomes that the organisation intends to achieve. One important factor in this respect is the issues that could pose an information security risk to the organisation's business objectives, purpose, information and information systems. It is also important to understand the organisation's risk culture and appetite in order to properly design, implement and integrate the ISMS within the organisation.

For the ISMS to be effective, appropriate and suitable to meet the organisation's objectives and purpose, it needs to determine the risks and opportunities. To achieve an accurate measure of what these risks are, the relevant internal and external issues and the requirement to be addressed in Section 3.1.2 needs to be identified and considered. This means for the ISMS to effectively manage these risks its design and implementation shall take account of these issues, relevant information security requirements and be "at one," in harmony, in sync with the context and business environment of the organisation. We shall in Chapter 4 address the requirements of ISO/IEC 27001 regarding risk management.

Understanding the business and its context is essential to enable the ISMS to work with and be embedded in the organisation and not function as a separate entity. This is important since the ISMS should be a business enabler, adding value to the business and minimizing the information security risks to help maximize its business opportunities.

3.1.2 Internal Issues and Context

All those internal issues and dependencies relevant to the purpose and objectives of the organisation achieving an effective ISMS need to be identified and taken into account. For example, the organisation needs to ask and consider if there may be internal standards, policies and procedures related to business processes or the management of operations and resources that are relevant. These may be issues, restrictions and dependencies that the ISMS needs to take into account when considering the information security risks and the implementation of controls for treating these risks. For example, restrictions on the type of technology that can be deployed because of internal procurement policy or a dependency from another department or business unit for a specific support or resource.

There may be internal IT infrastructure issues that may impact ISMS performance and effectiveness; again, these issues need to be taken into account in the ISMS risk assessment and treatment. The organisation may operate with internal service contracts between business units, and these may specify service delivery and availability issues and dependencies that the ISMS would need to address. There may be specific issues related to the internal workforce regarding organisation culture, dependencies on capabilities and competence of individuals for specific roles and awareness issues.

Understanding the internal environment is especially relevant in order to appropriately address the ISMS risks. There are internal issues and dependencies that have an impact on the implementation of the ISMS but also the ISMS itself will have a direct impact on business operations. Therefore, the better informed we are of what issues and dependencies are involved in the organisation's working environment and their effect on implementing an ISMS, the better the risk assessment and the better the decision making will be at arriving at the most effective way to treat the risks. Good risk decision making, the type and level of risk control needed and the cost and benefits of implementing these controls depends on good information being available and duly considered during the risk management process. Of course, this also applies to the external business environment in which the organisation operates.

3.1.3 External Issues and Context

All those external issues and dependencies relevant to the purpose and objectives of the organisation's ISMS need also to be identified and taken into account. This includes the following areas of risk and business dependency:

- External organisations relationships:
  - Customers, clients, consumers;
  - Business partners;
  - Supply chains;
  - Service providers;
  - IT vendors.
- External business processes;
- External resources:
  - Connections;
  - External workforce;
  - infrastructure elements.
- Market conditions and competition;
- New laws and regulations;
- Environment.

Dependency on outside resources, services and infrastructure can all have a negative impact on the organisation if, for example, there is an incident that causes disruption to the supply of these resources and services or results in damage to the infrastructure. Failures or interruptions, for example, in energy supply, telecommunication services or transportation services can have a major impact on organisations.

Those organisations supplying such resources and services are of course responsible for managing their own information and IT security risks. It is, however, the responsibility of the organisation availing itself such external resources and services to identify external risks in order to be able to appropriately manage their own business risks. This is why it is important to understand the organisation's activities, operations and processes and to identify the external issues and dependencies to be able to properly manage its risks. The more well-informed the organisation is about these issues the better able it is to make decisions of how to treat and manage its risks due to external dependencies.

3.2 Needs and Expectations

3.2.1 Interested Parties

Generally an ISMS interested party is any person, persons or organisations who may be affected by or involved in the development, implementation and operation of an organisation's ISMS. Some interested parties may not only be affected by or have an interest in the ISMS, but they also may have a legal right or claim to the ISMS such as a shareholder, stakeholder or investor.

Interested parties include employees of the organisation managing the ISMS, investors, business partners, suppliers, customers and regulatory bodies. Of course, all employees are interested parties of the ISMS in the sense that they all should be involved in the implementation and management of risk related to their own job function. Some employees, however, will have more involvement than others depending on their job function, but irrespective of this all should have a vested interest.

Implementing an ISMS should give confidence to all interested parties that the risks to the confidentiality, integrity and availability of the organisation's information is being properly and effectively managed.

3.2.2 Requirements Relevant to the ISMS

Once the interested parties have been identified, then the needs, expectations and requirements of these parties relevant to information security should be identified.

3.2.2.1 Stakeholders and Investors

Stakeholders and investors want a good return on investment from the ISMS development and implementation that they are funding and supporting. This means they want to see that the organisation has an ISMS that is effectively managing the risks and opportunities. Stakeholders are clearly interested in maximizing the business opportunities to be achieved by implementing the ISMS and at the same time minimising the information security risks and hence controlling any potential negative business impacts. Key to this is of course having a good understanding of the requirements of interested parties in order to be able to satisfy these requirements to

Case Study: Supply Chains

For those organisations deploying supply chains as part of their business operations, there are a number of issues and dependencies related to managing the information security risks. This is specifically of interest to those supply chains that involve one or more ISMS implementations as part of the chain. For example, in any supply chain scenario the organisation is dependent on a number of business activities and functions being integrated into the supply chain processes. Supply chain management needs to take into account the security aspects for the delivery of goods or services across the chain. The business systems in the chain need to form secure partnerships to ensure they can react and respond appropriately to attacks and incidents targeted at elements in the supply chain. Therefore, it is critical that all weaknesses and vulnerabilities in the supply chain are identified. Also a good knowledge of the issues involved is relevant to the risks and impacts to the chain.

Disruption in the supply chain can have a ripple effect across the entire chain, interrupting activities and processes that depend on each other and finally impacting the business of the end user organisation. A security disaster, incident, compromise or failure in one part of the chain could have an impact one of more other parts in the chain, subsequently affecting one or several organisations in the chain. For example, a "botnet" attack could ripple through the chain and have a disastrous operational and economic effects.

Of course the incident that caused the disruption in the chain may not be man made or IT related, but it might have been caused by an environment problem (e.g., a flood, tsunami or earthquake). One of the factors to be considered in this instance is whether organisations in the supply chain are geographically located in high-risk areas or locations prone to natural disasters.

It is therefore essential for organisations to look at both the business risks and the opportunities of deploying supply chains. They need to make sure they have factored in all the supply chain dependencies and issues into their information security risk assessment, and as a result of this assessment determine an effective set of information security controls to manage these risks. They need to consider the issues of contingency, continuity and availability of services during worst-case scenarios, be prepared for the unexpected and be able to adapt and build in reliance mechanisms into their business systems and processes.

manage the information security risks and to provide adequate protection of the organisation's information and its business.

3.2.2.2 Customers and Business Partners

Customers, clients and consumers want some confidence that their information is being protected (e.g., their PII is safe and secure and will not be leaked, stolen or access by those that are not authorized to see or use it).

Business partners want some assurance that the organisation is fulfilling its contractual obligations (e.g., to deliver a product or service as agreed) or that its commercial information is in safe hands and is not been leaked or shared with competitors.

3.2.2.3 Connectivity and Outsourcing/Cloud

External customers or partners want some assurance that the networked connections they have with the organisation will not be a channel for malware, hackers or other security breaches. This means that the organisation needs to fully address the technical requirements of information security incident management and consider some form of information sharing and reporting agreement on incidents (which is especially relevant to supply change management arrangements).

Many organisations are outsourcing their business activities and so are dependent on business partnerships to execute various functional and operation aspects of their business. Examples of this include facilities management, transportation services, call centres and back-office activities. It also includes the increasing use of outsourcing of IT services, for example, using cloud services.

The more dependent organisations are on the use of such outsourcing services, the more reliant they are on the external service provider managing its own risk.

3.2.2.4 Internal Technology Requirements

The organisation may have a number of internal business requirements regarding the use of IT. For example, an organisation may have a particular business reason to allow some employees to use bring your own

Case Study: Supply Chain Requirements

For those organisations deploying supply chains as part of their business operations, managing the risks need to consider requirements for:

- Stock control management;
- Service delivery management;
- Alternative sourcing arrangements;
- Business interruption/continuity plans
- Off-shore supplies.

devices (BYOD) at work. Such a requirement means there are a number of information security issues to be considered regarding what is acceptable use of BYOD on site and off site (at home and in public places) as well as the security of BYOD itself. The use of BYOD raise questions of protection of business information on a device that contains private information, compliance issues, technical support issues, incompatibility with other IT used by the organisation, introduction of malicious code onto the organisation's network via BYOD and other risks. One of the other requirements in the modern business environment is of course mobile devices, mobile networks and services. Mobiles devices may be BYOD or devices such as phones or tablets provided by and owned by the organisation itself. Mobile devices, networks and services all have information security risks related to them and so it is important that an organisation identifies the requirements for such IT capability, understands and manages the risks associated with such technology and takes account of its requirements for the use of such technology in its ISMS risk assessment. This includes dealing with issues of theft, what types of business transactions are allowed using such technology, whether remote comments to the organisation's servers are allowed, use of wireless networks off site, use of such technology in public places, introduction of malware, noncompliance with regulations and secure disposal of such devices.

In addition, more and more use is being made of cloud services, and again the organisation needs to carefully consider its reasons for using such services. These services can provide many business benefits, but the security aspects need also to be considered. Organisations are finding cloud services such as the use of software application services, data storage and management and web site hosting to be economically and operational beneficial. The information security risks, however, raise questions that the organisation needs to understand and be able to deal with to protect themselves. This includes the security provided by third parties to safeguard sensitive and commercial information, as well as any PII that is in the cloud, the compliance issues relating to the protection of PII, data segregation of the organisation's information with that of other organisations, physical location of the servers used to deliver the cloud services, date recovery in the case of a major disaster at the cloud server sites and the investigation of incidents at the server sites.

Another area the organisation needs to consider is its strategy and policy, and hence its requirements, regarding the use of social networking/media, by both the business and employees. There are many risks involved with the use of social networking/media sites that both the organisation and its employees need to be aware of. The organisation therefore needs to take account of its requirements for the use of social networking/media in its ISMS risk assessment. Its requirements need to be considered alongside the potential risks: accidental or innocent disclosure of company confidential information by employees, intentional disclosure of such information for the purpose of fraud, financial gain, or identity theft, as well as the disgruntled employee who wishes to damage the reputation of the organisation. Other risks include online abuse, stalking, access to and downloading of inapprOpriate material, social engineering attacks (persuading employees to give information away or to trick them into doing something they should not be doing) and the introduction of malware embedded in attachments.

The organisation will need to consider its requirements for the issue of acceptable use of all its IT resources and facilities by employees (e.g., use of BYOD, mobile devices and networks, cloud services, social media sites and so on), which can be included in a corporate Acceptable Use Policy.

3.2.3 Gathering Requirements Relevant to the ISMS

Information on the needs and expectations of interested parties for information security can be gathered in various ways and from different sources:

- Internal business objectives and requirements:
    - Corporate policy and objectives;
    - Heads of business units and departments, project owners, employees;
    - Risk owners.
- Requirements specified in contracts and service level agreements (SLAs):
    - Clients and customers will want the organisation to comply with any clauses dealing with information and IT security included in contracts and SLAs (e.g., this may be a requirement for 24x7 data management services and call centre services);
    - A client may specify in a contract specific requirements for reporting on incidents that my affect them.
- Compliance requirements defined in legislation and regulations:
    - There are many laws on the protection of personal data that detail specific information security requirements;
    - There are telecommunications regulations that specify information security requirements;
    - There are laws that relate to relate to the admissibility of electronic evidence that organisations should be aware of regarding the collection of evidence during a security incident.

- Meetings and interviews with internal and external interested parties;
- Internal management review meetings;
- Feedback from customers and business partners.

3.3 ISMS Scope

3.3.1 What to Consider and What to Include

Defining the ISMS scope is an important initial task in development of the ISMS. It is important to get the aspect right before the organisation spends it resources in carrying out a risk assessment. This might sound like an obvious thing to do, but defining the ISMS scope can present a number of challenges.

The scope needs to take account of the internal and external issues referred to in Section 3.1 (see ISO/IEC 27001:2013 4.1) and the requirements referred to in Section 3.2 (see ISO/IEC 27001:2013 4.2). In addition, the organisation needs to consider all relevant interfaces and dependencies between its business operations and those of other organisations.

The ISMS scope can cover all or part of the organisation, so we need to define its boundary in the case that the ISMS scope is only part of the organisation (see Figure 3.1). The size of the ISMS scope, whether it is all or just part of the organisation, is a management decision and depends on the organisation's business aims and objectives, as well as the motivation for its particular choice of ISMS scope.

The ISMS scope might cover the whole or part of an organisation, or an organisation might choose to implement several ISMS. The ISMS scope might be defined and cover the activities of a business unit or department, a project activity or a set of business functions or services.

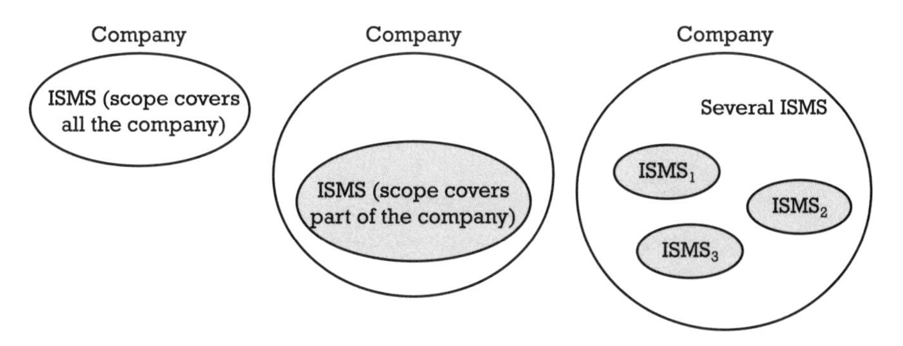

Figure 3.1 ISMS scope scenarios.

3.3.2 Object of ISMS Scope

Experience and feedback of existing implementations around the world inform us that organisations are selecting the ISMS route and a particular scope of application to achieve at least one of the following aims:

- Minimise the business impact and financial costs of security incidents in one or more business areas to:
    - Improve service delivery to customers;
    - Increase staff morale, productivity and effectiveness;
    - Reduce loss of contracts, sales, orders or profits;
    - Improve and enhance customer trust, confidence and relations, and meet contractual obligations of its customer(s), such as in the case of a company providing managed services to its clients;
    - Improve market position;
    - Reduce loss of assets and asset value;
    - Reduce e-risks;
    - Avoid legal/contractual penalties/liabilities;
    - Reduce breaches of operating controls.
- Reduce the number and frequency of incidents in one or more business areas:
    - Human errors;
    - Service interruptions;
    - Misuse/abuse of company resources;
    - Theft, fraud and other related crimes;
    - Malfunctions, system failures or system downtimes.
- Comply with a set of laws and regulations, such as the data privacy directive in the EU and the equivalent laws in the EU Member States, or comply with the Sarbanes Oxley (SoX) or HIPPA regulations in the US;
- As a market differentiator (e.g., a company that offers online—perhaps Internet banking service—or offline—say, media disposal or data recovery services);
- Enhance corporate value of the business and as a business enabler:
    - Maximise business opportunities and investments;
    - Provide more informed decision making about security risks;
    - Improve risk awareness, risk control and information security effectiveness for the organisation's sensitive and critical assets;
    - Facilitate an organisationwide information security risk culture.

- Publicly demonstrate that they are "fit for purpose" by means of a third-party independent audit, such as providing a particular line or type of customer-facing service (help desk and call centre facilities), as part of a supply chain, as a supplier of products or as a supplier of infrastructure supporting services (electronic signatures for secure payments and transactions).

3.3.3 Defining the ISMS Scope

Whatever the organisation decides its ISMS scope to be, it must ensure that it is well defined and covers all that it should cover in terms of requirements (ISO/IEC 27001:2013 4.2) and issues (ISO/IEC 27001:2013 4.1). The organisation should not exclude things from within the ISMS boundary if they seem to be difficult to deal with, as such exclusions could affect the organisation's ability or responsibility to provide information security that meets the security requirements determined by risk assessment and applicable legislative and regulatory requirements.

Depending on the scope of the ISMS, this might typically include:

- The staff and employees operationally involved in the ISMS area of work;
- The processes and services used within the ISMS area of work;
- The information and information systems necessary to carry out the business of the ISMS;
- The policies, procedures and documentation to be deployed necessary to carry out the business of the ISMS;
- The interfaces and connections to the ISMS;
- Supporting ICT infrastructure for the ISMS;
- The physical location(s) of the ISMS.

3.3.4 Scope Example

An example ISMS scope is a sales and customer service department of small manufacturing company. This group is just one of several departments in the company as shown in the in Figure 3.2.

The scope of the ISMS will typically involve:

- The sales and customer services personnel;

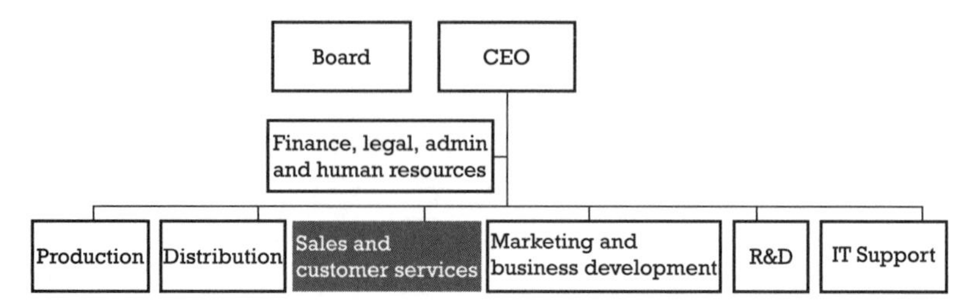

Figure 3.2 Organisational structure of a manufacturing company.

- The procedures used by the personnel:
    - Company polices and procedures;
    - Operating manual,
    - Security policies and procedures.
- Sales and customer service information:
    - Customer details;
    - Customer orders, invoices;
    - Internal company data;
    - Data from other companies.
- Use of business processes, applications and services:
    - Sales orders processes;
    - e-commerce;
    - Customer enquires processes;
    - IT services.
- Interfaces and dependencies:
    - Technical;
    - Nontechnical.
- Use of technology:
    - Desktop computers and other IT equipment;
    - Networks;
    - Telephones, mobiles, tablets.
- Physical location.

This example illustrates an ISMS scope that is only part of the travel company operations and the services it provides. Other scoping examples include:

- Internal services supplied by the ICT department;

- Call centre;
- Internal help desk support;
- An organisation's claims and payments department for its customers;
- ICT-based logistics support for a shipping or an airline company;
- The online hotel booking system for an international hotel chain;
- A research and development group of a software company.

3.3.5 External and Internal Connections

An important aspect of the ISMS scope is the connections that exist to and from the ISMS, whether these are with external customer or supplier systems or other parts of the same organisation. In the previous example, typical internal connections might be with the personnel department, those dealing with IT services and those dealing with physical security. External connections might typically be those with other banks, networks and ISP service providers. It is important to consider these connection interfaces since these interfaces are one of the ways that risks might present themselves to the ISMS. Management and control of the flows of information across these interfaces is important to protect the ISMS against these risks.

Managing these interfaces can be done in many ways through the use of contracts, SLAs and the use of operational procedures and technical controls. ISO/IEC 27002 (clause 6.2) provides a list of things to be considered when researching the inclusion of security aspects in contracts and SLAs.

CHAPTER

Contents

Managing the ISMS Risks

4.1 The Importance of Risk and Opportunity

4.1.1 Definition of Risk

There are several important changes in the area of risk management in the 2013 edition of ISO/IEC 27001 in comparison with the 2005 edition. One of these is the definition of risk, which has been made more generic and thus oriented toward a greater range of methods and techniques of assessing risks.

ISO/IEC 27001 (2013) is now aligned with the definition of risk used in ISO Guide 73 and ISO 31000, that is, *risk is the effect of uncertainty (on objectives)*. The elements of this definition can be interpreted in several ways, so "effect" can be thought of as the consequences or impact of an event occurring such as the consequences of a malware incident on an email server. The uncertainty could be that related to an event or incident (e.g., the likelihood of an unauthorised access attempt on a system containing sensitive information). The objectives could include the information security objectives found in ISO/IEC 27001 (i.e., preserving the confidentiality, integrity or availability of information). Another objective is that the ISMS is effective, suitable and adequate to meet the organisation's needs and requirements. So combining these elements we could have, as an example of a risk, the business impact of unauthorised access attempts on a system containing sensitive information would be detrimental to the organisation, not only damaging its reputation but also costing them $X in loss of revenue and potential legal penalties, and the likelihood or chances of this happening is Y out of Z.

4.1.2 Opportunity

An organisation needs to identify the opportunities that will contribute to achieving its strategic and business objectives. This can sometimes mean taking risks to gain access to these opportunities and sometimes reducing the risks to ensure these opportunities are available. The "effect" element in the definition of risk, given in Section 4.1.1, can also be interpreted as the deviation from the expected, that is, in terms of both positive as well as negative results, consequences or outcomes. This might be also expressed in terms of good or bad fortune, exposure to gain or loss, upside risk (favourable outcome) or downside risk (unfavourable outcome). This means that an organisation needs to create a balance between those negative risks that prevent opportunity and those positive risks that help to secure opportunities, and this will depend on the organisations risk attitude, tolerance and appetite.

4.1.3 Risk Attitude, Tolerance and Appetite

Should a business miss an opportunity to avoid risks or should the business accept/take risks to gain from the opportunity? This relates to the risk attitude of the organisation: are they risk averse (avoiding), risk neutral or risk affine (seeking) when it comes to going for business opportunities? The risk averse (risk avoiding) individual will try to reduce the uncertainty of a risk (i.e., when faced with either a choice where the loss or gain is uncertain or an option which is more certain but may result in lower gain, they would select the latter, more certain option). The risk seeker on the other hand takes a gamble on the uncertainty in the hope of a greater gain.

The risk attitude is a factor in determining their risk appetite to a situation or set of business choices. Risk appetite is the amount and type of risk that an organisation is willing to pursue, take, accept or retain to achieve its strategic and business objectives. Hence an organisation's risk appetite is linked to the pursuit of business opportunity. Different organisations have different risk appetites. An organisation with a low risk appetite wants to avoid the negative effects of the uncertainty of risk; it wants to pursue guaranteed gain from a business investment and not gamble on an uncertainty for a much higher gain. At the other end of the spectrum, risk-seeking organisations seek to gain higher rewards by selecting business opportunities and business innovations. Risk tolerance is the level or amount of risk an organisation is willing and able to accept.

The risk appetite and the levels of risk tolerance need to be defined by executive or senior management in conjunction with the board of directors and the CEO.

4.1.4 Information Security Risk Appetite and Tolerance

The information security risk appetite should be a part of the organisation's overall risk appetite. Defining the information security risk appetite, as with other risk decisions and criteria, should not therefore be an isolated activity but an integrated part of the organisation's overall strategy and objectives on risk. ISO/IEC 27001 emphasises the importance of the fact that management of risk comes from the top, with commitment, direction and decisions being made by executive/senior management.

Information security management should be seen as a driver for achieving the organisation's business strategy and objectives, and to add value to the organisation. Therefore, the risks related to information security need to be managed to achieve such business objectives and to satisfy internal and external requirements. The outcome and deliverables of the ISMS is therefore to provide effective protection of the organisation's information to achieve these business objectives and requirements.

Traditionally information security risks have been seen in a negative way, to protect against loss or damage, and not in the context of business opportunity to address the broader business strategy. However, as just mentioned, information security is not an isolated activity but one of the key drivers for an organisation to protect its business interests through protecting the information it uses to run and operate its business.

Of course there are information security risks related to legal, regulatory and compliance requirements, which the organisation needs to address, and because of the nature of these types of requirements sometimes there is no clear view of any underlying business opportunity. However, even in the cases of mandatory compliance there is always an aspect of business opportunity, which is especially the case as the approach to the management of risk is taken as part of business strategy. Conformity to the requirements of ISO/IEC 27001 demonstrates that the organisation is protecting its information and this includes addressing the risks using legal controls and protection. For example, the protection of personal data or personally identifiable information should start with the compliance with laws and regulations pertaining to the country of business. An example of this might be any service an organisation offers that involves the managing and processing of customer data, including customer PII. The service makes money for the organisation and so provides a business opportunity, but in pursuit of this opportunity the organisation needs to comply with the legal requirements that are relevant to its customer data and ensure adequate protection is provided in conformity with these requirements.

In considering information security appetite and tolerance, executive management decisions need to be made taking into account legal advice and counsel of the regulatory and legislative aspects.

4.1.5 ISMS Risks

Managing the effect of uncertainty on achieving the objectives, outcomes, targets and deliverables of the ISMS is the risk management task at hand. Are the information security risks managed to a tolerable level of acceptance so that the ISMS is effective, suitable and adequate for the organisation's needs and requirements? Does the organisation have effective, suitable and adequate ISMS policies, procedures, resources, processes, plans and programmes in place to manage the risks?

Things can change in an organisation and its business environment, which could have an effect on its risk profile: the business objectives and strategy may vary or change; operating conditions might change; market conditions might fluctuate—there may be economic downturns and upturns; the conditions surrounding and causing security incidents might change, with increased uncertainty regarding these incidents and greater threats and attack levels; there may be changes to infrastructure, internal and external; different business processes might be adopted; new and advanced technology may be deployed by the business. For example, the organisation might expand in a new business area and at the same introduce new technological solutions and deploy more staff as part of this expansion. The question is asked as to whether this expansion leads to changes that create new risks, or increase the level of current known risks, to the protection of its information. Is the ISMS still able to be effective, suitable and adequate to the organisation? This is what management should be asking. When changes occur, the organisation should reassess the risks and the opportunities associated with these changes as part of their continual improvement programme (i.e., addressing the risks to ensure that the ISMS remains effective, suitable and adequate for the organisation's needs and requirements).

4.2 Risk Management Process

4.2.1 Changes in the Process

As mentioned earlier, risk is the effect of uncertainty (on objectives). In the 2005 version of ISO/IEC 27001 there was mention of assets, threats and vulnerabilities in relation to the definition risk. So there was a need to identify the threats to assets and the vulnerabilities of the assets that the threats could exploit. These elements that made up part of the definition of risk in the 2005 no longer exist in the 2013 version of the standard. This does not mean, however, that an organisation that uses a risk assessment method using these elements needs to change its method. The 2013 definition is generic enough to accommodate a wide range of methods. What it does mean is that there is no mandatory requirement, in the 2013 edition, for an organisation to identify its assets, threats and vulnerabilities as part of its risk assessment, as was the case in the 2005 edition.

The overall process stages are the same: risk assessment and then risk treatment. Also stages in the risk assessment process are the same: risk identification, risk analysis and risk evaluation. What is different is in the detail of what constitutes identification, analysis and evaluation (i.e., there is no mention of assets, threats and vulnerabilities in the 2013 edition).

4.2.2 Risk Assessment

4.2.2.1 Process

It is entirely up to the organisation to define what information security risk assessment process it uses. This includes what methods and criteria it uses to identify and analyse its risks, and what acceptance criteria it uses to evaluate the risks it has identified. The 2013 version of ISO/IEC 27001 gives the organisation a lot of flexibility to define its own approach to suit its own business culture and to select the methods and techniques it wants to use as part of this approach. This approach needs to conform to the requirements defined in ISO/IEC 27001.

The assessment process should provide the organisation with an understanding of the risks they face. This is typically in terms of the risk causes and effects (consequences/impacts) and the likelihood of occurrence, the levels of risk and whether these levels are acceptable/nonacceptable in accordance with the acceptance criteria as defined by the organisation.

4.2.2.2 Acceptance Criteria

What is acceptable/tolerable or unacceptable/intolerable is defined by the organisation. The criteria for determining what acceptable levels of risk are will depend of the organisation's risk appetite and the significance of the risks to the organisation's business, objectives and values. The risks should also reflect both internal and external interests and requirements, relevant standards, laws and regulations.

4.2.2.3 Risk Owners

A very important requirement of the standard and the risk assessment process is to identify risk owners, that is, those individuals in the organisation that are accountable, and have the authority, for decisions regarding the management of the risks and ultimately regarding managing the risks. So, for example, an individual that is authorized and accountable for managing and controlling personal data of employees and staff would be the obvious person to be the risk owner of such information. Such an individual would be responsible, for example, for signing off the implementation of a set of controls for the protection this personal data and for approving the resulting level of residual risk and for the long-term accountability and risk management of this personal data. The same situation applies to all the other risk owners in the organisation.

4.2.2.4 Risk Identification

The process of identifying information security risks involves the identification and recognition of what could happen, which may have an impact on the organisation's ISMS and its ability to realize its business strategy and objectives. This includes events, incidents, potential breaches and compromises or any other situation that might have an impact on the organisation. Another aspect of this risk assessment stage is the identification of existing ISMS controls that are in place.

The identification process involves gathering together information regarding the risks obtained from individual and team interviews; feedback, opinions and views from stakeholders interested parties and experts; questionnaires/check-lists; records of information security incident management records and other historical data records such as audit reports and through various other methods and sources.

It is important that those involved in the information gathering and identification process recognize that risk perception can vary from one individual to another (e.g., what one individual perceives as a risk or a risk-related event may be differently perceived by another individual). Those in control of the identification process need to take account the possible variations in risk perceptions from both a human perspective as well as a corporate perspective.

The output of the identification process should be a list of risks, which will then be used during the risk analysis process (see next).

4.2.2.5 Risk Analysis

The analysis stage deals with determining the potential consequences/impacts of the risks, the likelihood of the risks occurring and the level of the risks. The level of risk is the combination of the consequences/impacts and the likelihoods. Figure 4.1 illustrates an example risk method for determining the level of risk given the potential impact (VL = very low through to VH = very high) and likelihood (VL = very low through to VH = very high). The risk levels in this example are given numerical values 1 = low risk through to 9 = very high risk. Clearly, this illustrates that on a sliding scale, the higher the impact and the higher the likelihood are, the higher the risk level will be.

During this stage of risk assessment, the causes and sources of risk need to be understood and considered, and subsequently the impacts and likelihood of these risks related to these causes need to be analysed. Of course there can be complex cause and effect scenarios: a set of different events add up to a specific risk. For example, an untrained member of staff who has little experience uses a badly written operational procedure, and this results in sensitive information being sent as an email attachment unprotected. The more experienced member of staff who has more experience and awareness of how to handle sensitive information may ignore the badly written procedure and protect the email attachment. This sequence of events and causes resulted in a risk for the organisation of its sensitive information.

An event may result in multiple consequences, which combine together to have an impact on several information security objectives. Of course, it might require a combination of causes to have an impact on the organisation. It is essential that the knock-on effects of causes and consequences are identified, including any cascading and cumulative consequences.

There are as many different ways and methods of doing the risk analysis as there are ways of doing risk identification. A group of staff carrying out a

Impact \ Likelihood	VL	L	M	H	VH
VH	5	6	7	8	9
H	4	5	6	7	8
M	3	4	5	6	7
L	2	3	4	5	6
VL	1	2	3	4	5

Figure 4.1 Example risk table.

brainstorming or interviewing exercise is certainly relevant for a risk identification, but not for calculating the impact or likelihood of the risks where a more systematic and analytic method is more useful and applicable. On the other hand, a cause-effect (scenario) analysis or a business impact analysis (BIA) is clearly relevant to the risk analysis with regard to consequences/impacts of the risk. Furthermore, root-cause analysis methods and "what-if" scenario methods cover both the consequence/impact and likelihood aspects of risk analysis.

In carrying out the risk analysis, it is important to understand the events that are likely have an impact on the types of information the organisation has: sensitive, critical, personal, restricted, private information and public information. The preservation of this information may be at risk, specifically regarding its confidentiality, integrity and availability. In addition, the protection of PII is also an issue. So, analysis of the events and causes that threaten this preservation is essential. This analysis can be carried out using different methods, including some of those mentioned earlier (e.g., cause-effect analysis, BIA), with input and feedback from various sources that are a legitimate and authorized interested party in this information (e.g., information owners, risk owners, top/senior management, users). In the same way that preserving business information is important, so are the critical business systems, process or services that are associated with the ISMS directly or indirectly. Such critical elements should be understood and their risks identified. Disruption, unavailability or continuity to these critical parts will have a high impact and detrimental consequences on the organisation and could have a dire affect on the organisation's operations. The complexity of the ISMS being risk assessed will depend on the resources (cost, time, competence) needed to do the risk analysis. This in part also depends on the scope of the ISMS: a well-focused scope oriented to a single business group, department, project or service, with well-defined noncomplex interfaces and dependencies is one level of complexity, whereas a scope that covers many operational sites, many parts of the of the business, many projects or many services with a complex set of interfaces and dependencies is a different, higher level of complexity.

4.2.2.6 Risk Table Scales

The organisation has the choice of which risk method it wants to use, the criteria for acceptance and the particular scales for assigning the levels of likelihood and the types and level of impact. In the example given in Figure 4.1, low, medium, high and very high are used for levels of impact and likelihood. Instead an organisation might decide to use a 3-point or 4-point scale, or a 6-point or higher scale. The organisation may wish to provide a more descriptive meaning to these points on the scale low, medium, high and very high. Figures 4.2 and 4.3 are examples to illustrate how this might look.

It is entirely up to the organisation whether it adopts a set of descriptors and the content of the descriptor. Whatever is adopted by the organisation should be aligned with its business objectives, culture and internal standards and methods it chooses to adopt (e.g., it would make sense to adopt the same descriptors it uses for other risk management activities it undertakes to achieve a corporate consistency, but again this is entirely up to the organisation to decide—these examples are merely generic illustrations to described the overall risk assessment process).

4.2.2.7 Risk Analysis of Existing ISMS Controls

Defining the level of risk should take account of existing ISMS controls and their suitability, adequacy and effectiveness. This involves considering what existing controls there are in place and their relevance to each of the risks that are identified and whether these controls are suitable and adequate to modify the risk to a tolerable level.

What level of effectiveness is provided by these existing controls? Are they user friendly or difficult to use? Do they work in the way they were intended to work? Are measurements available that will indicate the level of effectiveness?

In considering existing controls, one of the decision-making questions is whether it is best to improve the implementation of these existing controls to adequately modify the risks identified or whether it is best to implement a different set of controls.

Impact		Example description
Very high	Catastrophic	Threatens the survival of the company
High	Major	Major/significant losses, damage, harm
Medium	Moderate	Moderate losses, damage, harm
Low	Minor	Minor losses, damage, harm
Very low	Negligible	Little or no losses, damage, harm

Figure 4.2 Example impact scale descriptors.

Likelihood	Example description
Very high	Very likely (e.g.at least 4 in 5 (80%) chance)
High	Likely (e.g. at least 1 in 2 (50%) chance or more)
Medium	Likely (e.g. around at least 1 in 5 (20%) chance)
Low	Around 1 in 10 (10%) chance
Very low	Around 1 in 1000 (0.1%) chance or below

Figure 4.3 Example likelihood scale descriptors.

4.2.2.8 Risk Evaluation

The evaluation stage compares the results of the risk analysis stage, that is, the levels of risk with the organisation's acceptance criteria. For example, in the example in Figure 4.1, risk levels of 7 and above might be considered unacceptable and intolerable (dark grey area in Figure 4.4), whereas levels 1–3 are considered acceptable, tolerable and in general negligible (white area in Figure 4.4). On the other hand, risk levels 4–6 might require further analysis on a case-by-case basis, taking into account the business costs and benefits of risk treatment, and business opportunities against the business impacts, to determine whether any risks at these levels are acceptable or unacceptable (mid-grey area in Figure 4.4).

Figure 4.4 is merely one example of a risk acceptance heat map. The organisation needs to define its own dark-grey, mid-grey and white areas, and this will depend on the organisation's risk appetite. For example, an organisation might adopt the following heat map in Figure 4.5, as it has a lower risk acceptance.

Another organisation might adopt the heat map in Figure 4.6, as it has a higher risk acceptance.

The choice of risk criteria acceptance and risk appetite is entirely up to each organisation to define. Figures 4.4 to 4.6 are merely examples to illustrate the theory. Practical implementation is up to the organisation to select which risk criteria is suitable and adequate for its own business situation and circumstances.

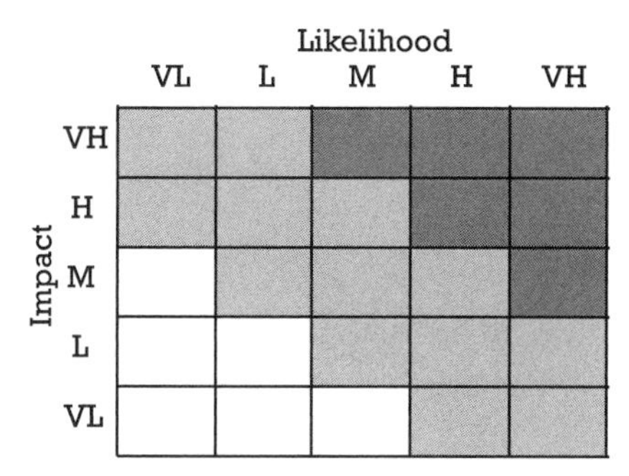

Figure 4.4 Example risk acceptance heat map (1).

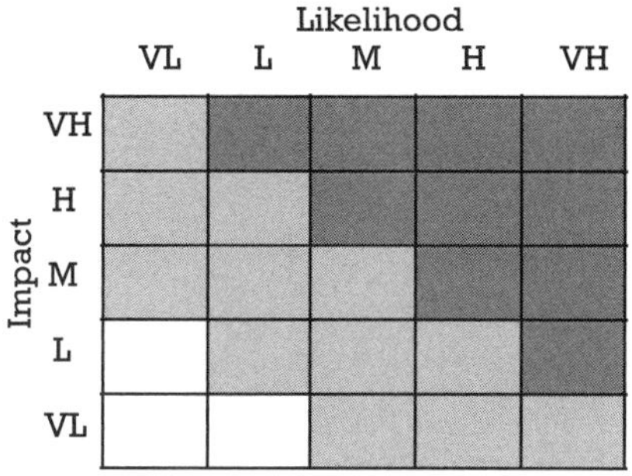

Figure 4.5 Example risk acceptance heat map (2).

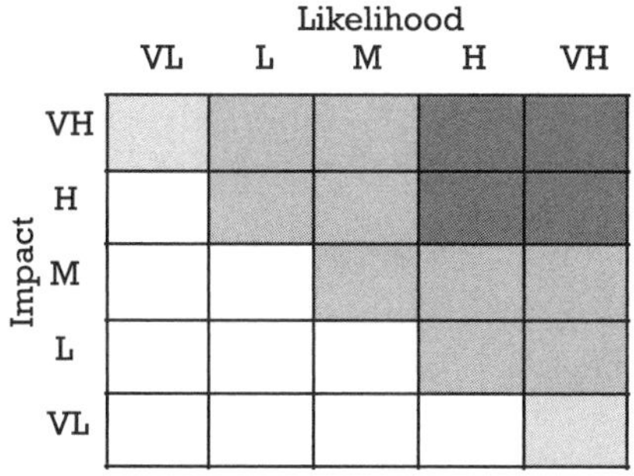

Figure 4.6 Example risk acceptance heat map (3).

4.2.2.9 Risk Evaluation Decision Making

So at this evaluation stage of the risk assessment, decisions need to be made as to whether risk treatment is needed to modify the identified risks to an acceptable level. The risks in the middle band in our example in Figure 4.2

may, after an analysis of the business costs and benefits of risk treatment, and business opportunities against the business impacts, fall into the unacceptable region (the upper band of the dark grey area) or they may stay in the middle band and, for example, may be classed as "for close monitoring and an early review/reassessment" to make sure the situation can be managed and does not change and creep into the upper band. Therefore, in the middle band, there is likely to be a scale of risks from the boundary (lower band) of almost being acceptable to the boundary of being not acceptable (upper band) and so the cost-benefits/business opportunities/business impacts need some comparative management decision making. One thing is certain—the cost of risk treatment needs to be kept in check to avoid the cost of modifying the risk becoming disproportionate to the business benefits and opportunities to be gained in engaging in a business activity that is associated with the cause of the risks. Therefore, management and risk owners need to be involved in this type of decision making, where the costs-benefits of risk treatment versus nonrisk treatment, balanced out with the business opportunities and impacts, are being considered in addition to any legal or regulatory issues related to the risks where legal council/advisors need to be consulted.

This decision making needs to consider the prioritization of the risks for treatment purposes. One of the major purposes of risk prioritization is to help in the allocation of resources to the treatment of risks and the implementation of ISMS controls to modify the risks, and of course the ongoing daily maintenance and upkeep of the ISMS.

Risks should be prioritized according to the most critical through to least critical ranking (e.g., very high impact, very high likelihood risks). The prioritization needs to take into account an assessment of the overall impact on the organisation and its critical systems and processes. High-impact risks regarding the commercial interests of the organisation need attention and generally fall into the category of critical, as might high-impact risks that might affect critical business systems or processes. Of course, the risk might be of high impact but only medium likelihood (falling into the mid-grey area); nevertheless, the risk relates to some piece of legislation (e.g., protection of personal data), in which case the priority criteria is in compliance with legislation.

4.2.3 Risk Treatment

After identifying the risks and prioritizing the risks for treatment, then the next process is to consider the treatment options that will modify the risks. The risk treatment process is typically an iterative process as illustrated in Figure 4.7.

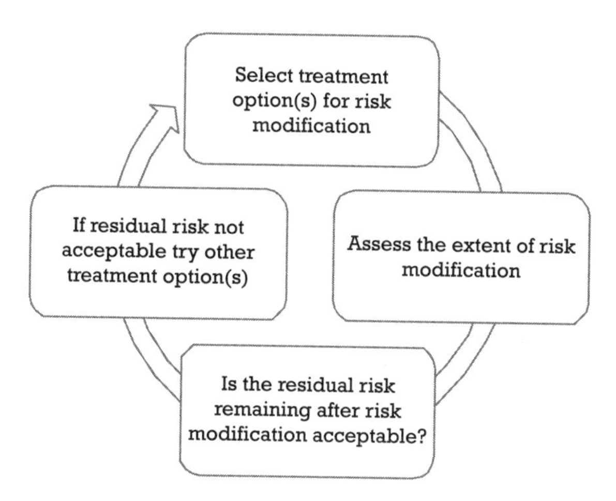

Figure 4.7 Risk treatment cycle.

Also included in the process is assessing how effective the risk treatment is at modifying the risk.

The selection process needs to be involve management as consideration needs to be given to the costs of treatment implementation against the business benefits that come from such treatment, taking into account the compliance requirements from legal, regulatory and contractual obligations. It might be the case that there are strong financial and business impact reasons regarding whether a treatment option is feasible. It might be the case, for example, that the impact on the organisation of damage to its reputation is very high but the likelihood of this happening is considered to be low to medium, which would place the risk into the middle band of risk, neither tolerable nor intolerable. This would mean further management discussion and enquiry into the economics of treatment.

The following are some typical types of treatment options:

- Risk avoidance (risk averse):
    - Terminate activities that currently cause the risk;
    - Postpone/cancel plans/projects/activities that could cause the risk.
- Risk taking (risk affine):
    - Pursuit of opportunity.
- Removing the source of the risk;
- Risk transfer:
    - Insurance;

  - Contracts;
  - Risk financing.
- Retaining the risk:
  - Knowingly and objectively accepting the risks through well-informed decisions.
- Modifying the risk by reducing/changing the likelihood;
- Modifying the risk by reducing the consequences/impact.

4.2.3.1 Risk Avoidance

Avoidance includes making the decision to not carry out, engage in or perform an activity that could cause the risk. Risk avoidance might involve changing the following:

- Company policy or strategy;
- Plans for location or relocating of business facilities;
- Methods, procedures or processes used;
- Operational conditions;
- Scope and objectives of new projects, developments or technology being used.

An example might be the relocation of the business to an area that is found to be less environmentally hazardous in order to not take on the liability that comes with the likelihood of these hazards occurring and interrupting, disrupting or destroying the business and its operations. It might be that the site for relocation is in a less dangerous area of a city to avoid serious incidents involving criminal activities and civil disorder. Avoiding this risk may seem a way of dealing with this problem but consideration needs to be given as to whether the organisation will be at a commercial disadvantage or will lose out on the benefits of not relocating.

Management thinking needs to weigh up the pros and cons of either avoiding the risk in order to take the opportunity of safer operating conditions or taking the risk in order to take the opportunity of other business benefits that may be afforded by the current location such as its customer base. The same avoidance strategy might taken regarding the risks of taking on new research projects, selecting backup sites for business continuity purposes or embracing new technologies such as digital accounting or online business.

4.2.3.2 Transferring the Risks

This treatment method involves employing another party, such as an insurance company, suppliers, venture capital business and external parties, to share some or all of the risk. Typically this could be by contract, insurance or by some form of hedging. Insurance is one type of risk transfer that uses contracts and in recent times e-risks insurance protection has started to emerge on world markets. Other means of transfer might involve the use of contract language that transfers the risk to another party without the use of an insurance premium. Such a contract might set limits of liability on contractors or service suppliers are very often transferred this way. Of course, in the financial world there are many other instruments of transfer for financially managing risks, such as taking offsetting positions in order to hedge against investment losses.

4.2.3.3 Reducing the Likelihood of the Risks

This involves using preventive, deterrent or detective types of control methods to reduce the likelihood of the risk happening in the first place or at least to make it more difficult for the risk to occur.
Information security polies, procedures and processes, user training and awareness programmes, removal of weaknesses, access control methods, and other measures can help to reduce the likelihood of risks. The actual effectiveness of these controls to reduce the likelihood will depend on the quality and effectiveness of the implementation and its operational/business use. For example, a control designed to prevent a security incident from being successful may not be effective in practice if the implementation is faulty or not properly done, and of course if users are not trained in the use of the control this also reduces the effectiveness of the control.

Procedures for instructing users on how to handle sensitive information can reduce the likelihood of errors and mistakes being made by users and the risk of such information being leaked to unauthorised users. Similarly a process for securely disposing of media used no longer can reduce the likelihood of information stored on that media falling into the hands of those unauthorised to see such information.

Controls that are able to protect against malware incidents help to reduce the likelihood of malware infections. If those same controls have the ability to recover from malware infections, they also help in reducing the impact of the infection if they are used in an effective and timely way.

Access controls can restrict the systems, applications, privileges and information that users can access, and the stronger these controls are the less likely it will be that unauthorised access will take place. This means the more difficult it is to penetrate through the access control system, the lower the likelihood of the risk happening—and it is much lower than if there were no access control system or the system was weak (i.e., easy to penetrate).

There are many controls that have the ability to reduce the likelihood of a risk happening and being realized. Of course, when a risk does materialize, then it can have a direct impact on the operational performance of the ISMS and so controls to reduce or minimise this impact also need to be considered. Some controls are able to work in combination to achieve both a reduction in likelihood and impact.

4.2.3.4 Reducing the Consequences of the Risks

If risk becomes a reality because, for example, the control methods for reducing the likelihood are ineffective, inadequate or unsuitable to prevent the risk, then controls are needed to reduce the consequences/impact of the risk. This includes using detective, responsive, reactive or recovery types of control methods to reduce the severity of the impact or loss caused by the risk.

Information security procedures and processes, information security incident management processes, business continuity and disaster recovery plans can all help to reduce the impact of risks.

Lack of management support, commitment and leadership for the ISMS; lack of management reviews and lack of or ineffective performance evaluation exercises can result in business impacts. The impact on the organisation failing to continually improve the suitability, adequacy and effectiveness of the ISMS will mean its ability to protect its sensitive and critical information will decrease and become ineffective. This will result in other impacts that will affect the organisation being able to carry out its business properly, productively and profitably.

An information security incident management process detects, investigates, responds to, reacts to and resolves actual, suspected or potential information security–related events. This process results in actions aimed to avoid or reduce the consequences/impact of the incident or reduce the likelihood of a further reoccurrences of the incident.

Some risks can cause severe disruption to an organisation's operations and productivity. Clearly, when such a disruption takes place the organisation wants to recover its business as soon as possible, as downtimes and lack of availability of systems and processes can have many impacts such as financial implications in terms of lost revenue, reduced productivity, lower quality levels, reduction in service levels, damage to its business reputation, harm to the trust and confidence its customers have in its business and the potential loss of customers who are not satisfied with loss of services as a result of the disruption.

The risks in some areas can have severe and dire business consequences and impacts. Lack of control regarding compliance with laws and regulations can leave the organisation with severe legal repercussions: fines, penalties, legal fees (e.g., failure to protect staff's and customers' personal data according to national legislation around the world).

In the area of physical security, computer rooms need to be protected against the risk of unauthorised access. Implementing physical access controls reduces the likelihood of risk of unauthorised access. However, if these controls are breached, then detective/response/reactive controls are needed to reduce the impact of the risk, (e.g., alarm systems, CCTV systems and so on).

4.2.3.5 Retaining/Accepting the Risks

This means the organisation is able to tolerate and retain the risk within its business risk management strategy by accepting the impact or loss when it occurs. Management must be fully knowledgeable and objective about this course of action and think carefully before adopting this type of treatment. At the end of the day, perhaps there is no other course of appropriate action to take. By default, all risks that are not avoided, transferred, reduced or transferred are retained.

Risk sharing within a group of individuals or companies is also a risk retention option. This involves spreading the risk across the group by transferring the risk liabilities and losses to all those members involved in the group. This is not the transfer by insurance option.

When the risks are small, acceptance or retention is a viable and financial strategic option. Of course, we need to bear in mind what might be small to one organisation may not be small to another in terms of the costs and financial impact they are able to sustain. Acceptance of risk is especially the case where the cost of reducing or transferring the risk could add up in the end to be greater in the long run than the total losses sustained.

4.2.4 Determine the Controls

After the risk treatment options have been selected, a set of controls need to be determined to implement these options—see Chapter 6 on controls to modify risks. A gap analysis is then carried out comparing this set of controls with the controls set out in Annex A. This check aims to make sure that no control areas are overlooked.

This process appears in the 2013 edition of ISO/IEC 27001 but not the 2005 edition. The process that was defined in the 2005 selected controls from Annex A, which is not the case in the 2013 edition—see Chapter 6, Section 6.1.1.

4.2.5 Statement of Applicability

The results of the control determination and gap analysis processes are brought together in what is called a Statement of Applicability (SoA). This provides a useful record of the results of risk assessment and risk treatment processes. The SoA is also a key piece of documentation for ISMS certification audits (see Chapter 10).

4.2.6 Risk Treatment Plan

A risk treatment plan should be produced that details how the risk treatment options that have been chosen will be implemented by the organisation. This plan should provide the reasons that the specific selection of treatment options has been chosen and what the expected benefits that the organisation expects to gain are. The plan should detail the priorities for and actions to be taken to implement the treatment options, as well as legal and regulatory measures to be taken, the resources required for the implementation, and the schedule and action delivery dates for implementation. Those risk owners who are accountable for approving the plan need to be identified, as well as those responsible for implementing the plan. In addition, performance metrics and measurements need to be determined, and a reporting mechanism needs to be specified.

4.2.7 Risk Owners' Duties

The risk owners that are accountable and have the authority for decisions regarding the management of the risks and ultimately regarding actually managing the risks need to approve the residual levels of risk as well as approve and sign off on the risk treatment plan.

4.3 Ongoing Reassessment of Risk

4.3.1 Risk Reviews and Reassessments

A risk assessment is a snapshot in time of the risks the organisation faces when the assessment is carried out. The conditions, circumstances and other factors that were applicable to the risks that were identified during any assessment will change at some point in time; this is the nature of everything in the world around us—nothing is permanent.

Organisations should expect that changes will happen, and that this will change their risk profile. For this reason, a reassessment of the risks needs to be undertaken at regular intervals. What is a regular interval? This is up to the organisation to determine; however, typically this could be every six to nine months or sooner if there are known or planned changes or there is a major incident or event that has caused disruptive risks to the organisation. Some organisations set their reassessment interval at three to four months because of the nature of their business environment and the dynamic business environment they operate in.

4.3.2 Risk Monitoring

As conditions and factors that will have an effect on the risk profile of an organisation will change, vary and fluctuate, ongoing monitoring is essential to keep up to date and for continual improvement of the ISMS to ensure suitable, adequate and effective protection. In Chapter 9 we shall look into the topic of monitoring and measurements in more detail. At this point we shall only provide an overview of the importance of risk monitoring.

4.3.3 Updating the Risk Treatment

After reassessing the risks, there will be a need to reassess the ISMS risk treatment options. This needs to take into consideration whether to take corrective action to improve the existing set of controls, to implement a different set of controls and select additional treatment options. This will involve producing an updated risk treatment plan.

CHAPTER

5

Contents

ISMS Leadership and Support

5.1 Management Policy

An important aspect of conformance to the requirements of ISO/IEC 27001 and of achieving a successful ISMS development and implementation and ongoing management is that such task should be driven and led from the top, by top management.

Top management should start by defining an appropriate information security policy for the ISMS. The policy should be a clear management statement of its intentions, objectives and goals regarding information security and the protection of its information systems. This policy should reflect top management commitment and support for the ISMS to satisfy the requirements in Section 4.1 (ISO/IEC 27001:2013) and address the issues in Section 4.2 (ISO/IEC 27001:2013). It should be a directive from above that typically should address at least the following:

- The scope of information security, its importance to the business, and clarity about what the business information security objectives are (e.g., regarding the protection of the confidentiality, integrity and availability of its information);
- The need for staff awareness: staff should be aware of their duties and responsibilities regarding the risks (e.g., their responsibility to handle and process sensitive company information in way that protects it from compromise);
- What is acceptable and not acceptable with regard to behaviour and use of its resources (e.g., acceptable use of the company email system);
- Its obligations to carry out its business in compliance with the laws and regulations, contractual obligations, best practices and standards that staff also need to comply with (e.g., compliance with laws on copyright, data privacy/protection and computer use/misuse/abuse);
- Reference to any other documents that staff need to be aware of and comply with (e.g., more detailed security policies and procedures as well as any other relevant proceedings not directly related to security). This could be industry-specific policies such as those businesses that need to deal with environmental issues, aspects of health care, production of pharmaceutical products or food safety.

This management information security policy should be written in a way that the style and content are independent of any particular skill, process or technical knowledge. For example, the content should be understandable by someone that is not an IT specialist, someone not trained in company finances or legal affairs or does not have human resources skills. In other words, it should state information security objectives that are generally understood by all staff not just people with highly technical backgrounds or certain professional qualifications or skills.

5.1.1 Approval, Communication and Awareness

This management policy needs to be approved and signed by the CEO (or someone of similar management authority and accountable status), since the aim is to indicate management commitment and support.

The policy needs to be communicated across the organisation to all staff and interested parties. This could be in paper form or by electronic means or both. Some organisations display their policies on the walls of offices, computer rooms and other areas to ensure they are continually accessible and visible. Other organisations resort to using ICT to distribute and have available their policies and procedures via their internal network. Others may choose to distribute it in paper only for staff to keep at their own place of work. Whatever the method used, the policy should not be hidden away and forgotten. Staff need to read, understand, and refresh their memories every so often about its contents and what it says about their specific information security responsibilities and duties.

5.1.2 Policy Review

This policy needs to be reviewed and updated as necessary to take account of the changing nature of the information security risk environment and evolving organisational developments and changes. There needs to be a review process for maintaining this policy, as is the case with all policies and procedures, as part of the ISMS continual improvement process.

5.1.3 Management Policy Sets the Scene

This management policy is a high-level policy that "sets the scene," and typically there will more detailed policies (that will be covered in Chapter 7), which will give more specific rules and instructions on the implementation of information security protection. For example, policies on access control cover the rules of access to different organisational resources and facilities such email servers, databases, network services, applications as well as physical access to buildings, offices, rooms and storage equipment.

5.2 Leadership

Demonstrating leadership in regard to the ISMS is a core aspect of the ISO/IEC 27001 standard. It is essential that top management provide the appropriate level of leadership in terms of direction, authority, policy, governance and organisation. Good leadership defines the business purpose of information security, creates the mission statement, sets the strategy, provides staff focus on what is important with regard to the information security for the business and what the priorities are, motivates and inspires confidence and trust in the workforce that it is committed to protecting the business and nurtures security culture and security skills.

Good ISMS leadership is needed to build a team that will successfully take forward the implementation of the ISMS, which will empower and motivate staff to be proactive followers and supporters in helping to protect the organisation. A good ISMS leader will be passionate about being successful in managing the information security risks the organisation faces. ISMS leadership should strive to inspire others to see information security as a business enabler, with the vision of turning information security risks into a business opportunity.

Leadership is different than management—the former motivates and inspires, creates the vision and points people in the right direction, while the latter administers, controls and follows the vision and organizes people. Both ISMS leadership and ISMS management together achieve an effective, robust, resilient ISMS. Leadership will be the champion of the ISMS, and management will control and manage the ISMS.

5.3 Roles and Responsibilities

Day-to-day working and operational activities functioning effectively, and the proper management of staff, at all levels throughout the organisation, can contribute to an effective information security business environment. In part, this requires a good information security culture within the organisation to be in place, with appropriate awareness and understanding of the problems of information security risks and clear lines of responsibility and accountability.

It is essential that roles and responsibilities for protecting specific types of information or information systems or for carrying out specific information security–related processes are clearly defined and allocated. For example:

- The owner of an information system should be given the information security responsibility and accountability for that system (of course, these owners may delegate that day-to-day implementation of security to another individual or to a service provider, but they remain ultimately accountable for the protection of the system and the management of the information security risks);
- Personal data manager;
- Business owner—a specific department/group (ensures implementation of policy and procedures, defines information usage and classification for information in their custody, allocates information custodians, defines access roles and privileges, conducts staff training and awareness and provides protection of personal data under their control);
- Chief information security officer (CISO);
- Information security incident response team;
- Business continuity manager;
- Internal auditors;

- Human resource manager;
- IT services manager (IT service management, IT disaster recovery, involvement in incident management);
- IT and network administrators/managers (network management, secure network technologies, involvement in incident management);
- Authorised users of information systems.

In addition, all staff will have general responsibility of information security related to their day-to-day work. For example, reporting of unusual or suspicious behaviour either related to their use of IT, network services or related to other staff or visitors. Also all individual staff needs to be aware of their responsibility for keeping their passwords and other types of access codes secure, to ensure that they are using organisational resources in accordance with the acceptable usage policy (e.g., rules for using email for sending file attachments).

5.4 Resources

As with any successful business venture, it is important to have the right types of resources for the jobs that need to be done. Having staff with the right competence to do a job properly, efficiently and effectively is key to the overall success of the business. If it is a technical job, then the staff involved need to have the right level of knowledge and skill to handle technical requirements of job at hand, to resolve technical problems and to be able to use techniques, methods, equipment and procedures relevant to the technical area in question. If it's a customer service job, then the staff involved must have the relevant skills needed to deal with customers (e.g., they are able to listen and respond effectively to customer's questions and queries, they are able to satisfactorily resolve customer queries and problems, follow up on feedback from customers and generally be able to meet the expectations of the organisation's customers).

Management needs to ensure that for specific information security tasks, it has the right people, with the right skills and knowledge and experience. This can mean recruiting people who have the right existing skills and experience or recruiting people and providing a training programme for them to develop the right skills and experience. In addition, all staff working in the field of information security, whether those with experience or those in training, need to keep up to date as the issues and risks in information security continually evolve, as does technology and business practices.

Every organisation strives to have human resources with certain core competence, and many organisations seek staff with information security skills. The market is expanding, becoming more buoyant and becoming highly competitive. For many years, organisations have recruited information security personnel with hands-on experience, practical knowledge and appropriate references. Even though this is still the general basis of recruitment, more and more organisations have started to request applicants have market proven professional qualifications, personal certifications and in some cases university qualifications or a combination of both.

The current education and training market provides certified qualifications, for example, in areas such as:

- Information security auditing;
- Information security management;
- Risk management;
- Information assurance;
- Governance;
- IT security;
- Physical security.

Many universities around the world are now offering information security and IT security modules as part of Bachelor of Science degrees, some are offering Masters of Science degrees in information security, and some run PhD programmes in information security research.

The market for information security education and training is growing rapidly, which is a reflection of the increased awareness of the risks and the importance of protecting information. This trend means that in the near future, organisations will have access to a greater number of human resources with professional qualifications to complement the practical on-the-job experience and skills needed from information security practitioners.

5.5 Training and Awareness

The organisation needs to ensure that staff are aware of information security risks and have sufficient understanding to support the organisation's information security policy to undertake their normal work functions and tasks. Staff should be trained in the use of information security policies and procedures, security controls applicable to their job function and the correct use of IT (e.g., log in procedures, keeping passwords safe, appropriate use of IT).

5.5.1 When Should Training Take Place?

- Induction training for new staff upon joining the organisation. This should cover the company's information security policies, procedures and routine practices, whom to contact for help and support regarding information security matters and whom to report security problems to, initial familiarisation with the common types of risk malware, hacking, protection of commercially sensitive information and the protection of personal data, fraud, use of email and so on.
- On-the-job training providing specifically tailored instructions on information security suited to the individual's job function.
- Annual (or more frequent) refresher training to keep staff up to date with new developments and to provide organisationwide reminders or more immediate remedial training as the result of a security incident or an emerging risk.

5.5.2 Training Methods

An organisation can deploy a variety of methods to deliver effective information security awareness and training to its staff. The methods that an organisation selects will depend on the business culture and its operational needs. Therefore, an information security awareness and training programme should be tailored to the specifics of the organisation. You should alternate between different methods, perhaps introducing an element of motivational instruction together with practical interactivity.

- Classroom-based training can be highly interactive—such training can vary from half-day/one-day induction/beginners training course, through to various intermediate/advance three-to-five-day training courses covering a range of specific topics.
- Computer-based/online web-based training and awareness is a good method for reinforcing information security principles and specific topics. Such training can be delivered as a set of modules, interactive or noninteractive, and be accessible to staff at a time and place convenient to the individual.

- Seminars, workshops, round-table discussions and presentations are especially well suited for introducing new subject matter and for organisations with multiple sites;
- Videos are also an effective way to provide training on various topics.
- Posters provide visual reinforcement of information security principles and specific topics.
- On-the-job/des-top training is available.
- Internal emails can be used to remind, reinforce and provide updates on organisational policies and procedures.

5.5.3 ISMS-Related Topics

Implementing the requirements of ISO/IEC 27001 covers many different tasks, activities and processes that need to be carried out, and these are associated with a number of specific topics that information security professionals, practitioners and other staff will need to have knowledge of and develop experience in, depending on their particular job function. For example, an internal ISMS auditor will require knowledge of the auditing process and methods, whereas an ISMS risk manager would need knowledge and skills of the principles of risk management. These include:

- Principles of risk management (e.g., as found in ISO/IEC 27005, NIST SP 800-30, -37, and -39):
  - Risk assessment;
  - Risk treatment.
- Information security controls:
  - Generic controls (e.g., as found in ISO/IEC 27002, NIST SP 800 ...);
  - Sector-specific controls (e.g., as found in ISO/IEC 27010, 27011, 27015, 27017, 27018, 27019).
- Performance evaluations:
  - Measuring and monitoring methods and techniques (e.g., as found in ISO/IEC 27004);
  - ISMS auditing process and methods (e.g., as found in ISO 19011 and ISO/IEC 27007).
- Certification and auditing:
  - Certification (e.g., as found in ISO/IEC 27006, ISO 17021);

- Auditing (e.g., as found in ISO/IEC 27007, ISO 19011).

- Legislation and regulations related to information security and privacy:
    - National and regional laws and directives;
    - Standards covering privacy (e.g., as found in ISO/IEC 29100, 29134, 29190).
- Other specific security-related processes:
    - Incident handling (e.g., as found in ISO/IEC 27035, NIST SP 800-61);
    - Business continuity (e.g., as found in ISO 22301, ISO/IEC 27031);
    - Operational resilience (e.g., as found in BS 16000).
- Specific IT security controls and mechanisms:
    - Network security (e.g., as found in ISO/IEC 27033);
    - Applications security (e.g., as found in ISO/IEC 27034);
    - Malware (e.g. as found in NIST SP 800-83);
    - Firewalls, IDS, IPS;
    - Access control.

CHAPTER

6

Contents

Controls to Modify the Risks

6.1 Determining the Controls

6.1.1 Control Framework

As mentioned in Chapter 4 ("Managing the Risks") of this book, the 2013 version of ISO/IEC 27001 retains the general principles of risk treatment specified in the 2005 version—determine an option or options for risk treatment; identify, determine or select controls; produce a statement of applicability; produce a risk treatment plan; gain approval and sign-off of plan and residual risks. There is, however, a major specific difference, and that concerns the security controls used in the risk treatment process between the 2005 version and 2013 version of the standard. In the 2005 version, the controls were selected from Annex A of the standard; however, in the 2013 version controls are not selected from Annex A, but determined (Section 6.1.3) by design or identified from other sources (e.g., codes of practice, industry standards, sector-specific standards). This change in emphasis from the selection of Annex A controls to determining controls from any source provides greater flexibility and choice for organisations. Annex A is still retained in the standard, but its function and purpose has changed. Its purpose in the 2013 edition of the standard is as a benchmark to check that the controls that have been determined, by design or identified from other sources, to cover as a minimum the set of baseline controls addressed in Annex A. If there are controls in Annex A that are not included in the set of determined controls, then justification why they have not been identified for implementation needs to be given.

6.1.2 Process of Determining a Control Set

Controls to implement the risk treatment option are selected based on their capability of modifying risks. For example, a control or a set of controls in combination may be used to prevent, detect, reduce or correct the undesirable effects of an information security incident, breach or compromise. Controls that are used to prevent an information security incident include those controls that reduce the likelihood of the incident occurring in the first place. For example, installation of malware detection software can be used to prevent infection of IT systems and subsequent loss, damage or destruction of information. A well-written and properly deployed disaster recovery plan and procedure can reduce the consequences/impact on the organisation in the event of a disaster.

Some controls are able to detect indications that an information security incident or attack is about to occur. If such detected indications are effectively acted upon in a timely manner, they can reduce the potential impact of the attack. In some cases, they can stop the attack from occurring in the first place.

If the knowledge gained from an employee information security awareness and training programme is properly put into practice by employees in their everyday jobs, this contributes to reducing the likelihood and impact of an attack. For example, if employees know when, how and why they need to report about potential events that might be a precursor or an indicator of an information security attack or incident, then this helps in the analysis and response, which subsequently can help in reducing the potential impact of the attack. Equally if employees know what to do to protect the sensitive information they are processing by following a company procedure that gives instructions of how to protect such information, then they are contributing to reducing the likelihood that sensitive information will be leaked or stolen.

So in determining the controls needed to (a) reduce the consequences and/or (b) the likelihood of the risk, account needs to be taken of the extent that the control:

- Reduces the likelihood of a risk-related ISMS attack, incident or event occurring (which can mean reducing or removing the weaknesses or vulnerabilities in the business systems or processes, improving the effectiveness of a control to prevent an information security incident from being realized);
- Reduces the consequences or impacts of an attack, incident or event (which can mean implementing controls that will react to an information security incident that is happening or is expected to happen to limit or reduce the consequences of the incident or correcting the effects and impacts of an information security incident);
- Detects and responds to precursors and indicators of an attack, incident or event;
- Implements preventative actions against the future occurrence of an attack, incident or event;
- Implements corrective actions to remove the cause of a past attack, incident or event.

6.1.3 Existing Control Sets

In determining which controls are needed to modify the risks identified in the risk assessment process, it is also important to take into account any existing controls. Any system of controls needs to work together, and so it is necessary to understand what controls already exist and what role and effect they have in modifying the risk. It is also important to check whether any new controls that are planned for implementation conflict with the existing set, whether there is redundancy or how effectively will they work together and interwork.

6.2 System of Controls

6.2.1 Control Framework

Controls are determined as part of the risk treatment process on the basis that they are able to modify the risks identified in the risk assessment process. Controls can be of many forms such as policies and procedures, processes, plans or techniques. It is important to bear in mind that ISO/

IEC 27001 is a management system standard and the object of protection is information, so the controls are management-oriented controls not IT controls, as this standard is not an IT security standard.

Figure 6.1 shows a typical system of controls.

6.2.2 System of Controls

6.2.2.1 Policy and Procedural Controls

Examples of policy and procedural controls include:

- Information security policy;
- Acceptable use policy;
- Information handling and transfer policy and procedures;
- Access control policy and procedures;
- Backup procedures;
- Secure operating procedures;
- Network and communication security policies;
- System change control procedures (and processes);
- Recruitment processes;
- Information security incident management;
- Procedures related to operational and application software;
- Procedures relating to working in secure areas.

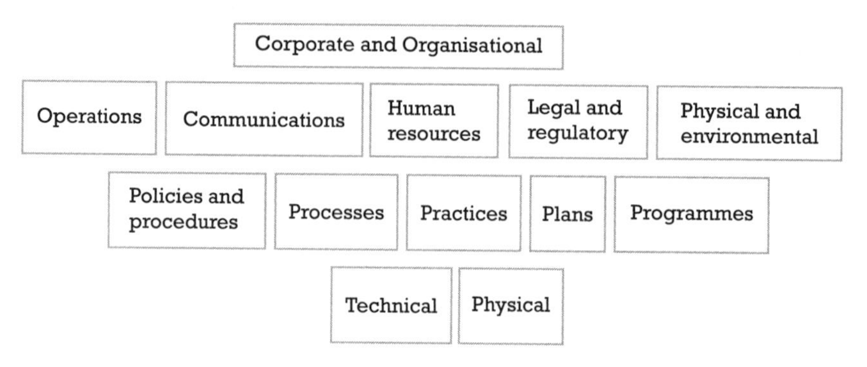

Figure 6.1 Control framework.

6.2.2.2 Process Controls

Examples of process controls include:

- Assignment of organisational roles, responsibilities and authorities relevant to ISMS;
- Risk assessment and treatment processes;
- Document management processes;
- Processes for determining competence related to human resources and information security performance;
- Processes for determining resources for ISMS;
- Management authorization processes;
- Internal and external communications;
- Monitoring and performance evaluation processes;
- Internal ISMS audits;
- Management reviews;
- Recruitment processes;
- Disciplinary processes;
- Processes of changing or terminating employment;
- User access registration and deregistration processes;
- Formal user access provisioning processes;
- Information security incident management;
- Backup processes;
- Outsourced activities.

6.2.2.3 Practices

Examples of control practices include:

- Organisational working practices;
- Operational practices;
- Employment practices;

- Legal practices;
- Sector-specific best practices.

6.2.2.4 Programmes

Examples of control programmes include:

- Internal ISMS audit programmes;
- Recruitment programmes;
- Awareness and training programmes.

6.2.2.5 Legal controls

Examples of legal controls include:

- Contracts containing security requirements;
- Service level agreements;
- Laws and regulations, directives.

6.2.2.6 Plans

Examples of plans include:

- Risk assessment exercises;
- Internal ISMS audits;
- Capacity planning;
- Disaster recover;
- Business continuity.

6.3 Policies and Procedures

6.3.1 General

Information security policies can either be a high level of policy or a more in-depth level of policy. ISO/IEC 27001 (clause 5.2) specifically refers to top management policy; this is typically a high-level policy. An organisation may also have policies and procedures at different levels of detail targeted at specific topics such as (a) mobile device policies, (b) access control policies, (c) clear desk policies, (d) backup policies and (e) policies on the use of cryptographic controls.

The top management information security policy (ISO/IEC 27001 clause 5.2) is intended to be a high-level directive and mission statement appropriate to the organisation and its business. Such a policy includes the information security objectives, a commitment to satisfy the requirements of interested parties (ISO/IEC 27001 clause 4.1 and 4.2) and an undertaking to continual improvement of the ISMS (ISO/IEC 27001 clause 10).

This information security policy should be written in a way that the style and content are independent of any particular skill, specialist or technical knowledge; it should also be communicated to all in the organisation and to other interested parties as and where appropriate. The content should be understandable by someone that is not a specialist (e.g., someone who is not trained in company finances or does not have human resources skills). In other words, it should state information security objectives that are generally understood by all staff, not just those with a highly technical background or certain professional qualifications or skills.

The length of the documented policy (number of pages) is not specified in the standard. Some organisations produce such policies of between one and three pages, others may decide to produce a policy that is greater in length and volume of text. It is up to the organisation to decide the length, volume and detail of text in its policy commensurate with its business objectives, strategy and culture. What is more important is that it contains enough information to convey in a clear way the objectives and ideas referred to earlier.

The broader aspects of information security policy at all levels of detail should be to:

- Provide a clear definition of the scope of information security, emphasize its importance to the business, and clarify what the business information security objectives are (e.g., regarding the confidentiality, integrity and availability of its information);
- Make staff aware of their duties and responsibilities regarding protecting the information the organisation has control over (e.g., their responsibility to handle and process company restricted and confidence in way that protect it from compromise);
- Set out what is acceptable and not acceptable with regard to behaviour and use of its resources (e.g., acceptable use of the company email system);

- Make clear its obligations to carry out its business in compliance with the laws and regulations, contractual obligations, best practices and standards with which staff also need to comply (e.g., compliance with laws on copyright, data privacy/protection and computer use/misuse/abuse);
- Give reference to any other policy documents that staff need to be aware of and comply with (e.g., more detailed security policies and procedures as well as any other relevant proceedings not necessarily directly related to security). This could be industry-specific policies such as those businesses that need to deal with environmental issues, aspects of health care, production of pharmaceutical products, or food safety.

6.3.2 Approval, Communications and Awareness

Information security policies and procedures need to be approved and signed by someone of management standing, who has the authority and is of accountable status. This achieves one of the aims of information security policy (i.e., to indicate management commitment and support).

Information security policies and procedures should be communicated to all staff. Some organisations display their policies on the walls of offices, computer rooms and other areas to ensure they are continually accessible and visible. Many organisations today resort to using electronic means to distribute and have available their policies and procedures via their internal network. Others choose to distribute in paper only for staff to keep at their own place of work. Whatever the method used the policy should not be hidden away and forgotten.

Staff must not only be aware of the policy, but also understand its content and what is says about their specific information security responsibilities and duties. Staff must read, understand and refresh their understanding on a regular basis the contents of policies and procedures.

6.3.3 Review

Like most things in business, there will be changes such as how the organisation operates and how it does its business. Therefore, changes should be reflected in the information security policy. A review should be made of the policy whenever it is most appropriate, possibly after identifying changes to the business, after an incident or as part of a regular management review on its ISMS (ISO/IEC 27001 clause 9.3).

6.4 Example controls

6.4.1 Overview

This section provides some examples of policy and procedural controls. These examples provide details of what might typically be included in the policy and procedural controls. The examples are not the finished product of such controls; they are merely examples and should be read and considered as such. The finished product would need to be customized to the organisation itself and its requirements and objectives. In some cases (e.g., the acceptable use policy), the organisation would need to have the wording of such policies and procedures be produced by management and checked by legal experts and human resources, as the content and wording of such policies will depend upon the jurisdiction of the country in which they are used applied. Additional information on operational controls, such as backup processes, information security incident management procedures and business continuity plans, will be given in Chapter 8.

6.4.2 Acceptable Use Policy

6.4.2.1 Basic Idea

Many organisations do allow their staff to have access to the Internet for personal use as a goodwill gesture to improve employee relations. In such cases, as well as the general use of an organisation's assets and resources for business, it is a very good idea to put in place an acceptable use policy (AUP) with regard to the use of the organisation's resources—in particular, its IT systems that are deployed for email, Internet access and services. The AUP policy is addressed in ISO/IEC 27001 A.8.1.3 under Assessment Management.

The basic intent of such a policy would be to protect the organisation, its staff and management, its business partners and its customers and suppliers from individuals involved in any illegal, incriminating and/or damaging actions, whether knowingly or unknowingly. The consequences of such actions could expose the organisation and its staff, business partners, customers and suppliers to a range of risks, impacts and possibly civil and criminal legal proceedings, actions and lawsuits.

The following is a word of caution regarding the content of the AUP. An organisation that does not explicitly state what is not acceptable and what is prohibited might face some staff relations issues. For example, a member of staff is dismissed for unacceptable use of the Internet for personal use. The organisation faces the risk that the member of staff takes legal action for unfair dismissal and is claiming compensation since the activities they were engaged in were not expressly prohibited. It is always good practice to make it clear to staff that access to the Internet and email facilities for personal use is a privilege and not a right. Legal and human resource advice on the content of such a policy needs to be sought.

6.4.2.2 General Policy Provisions

In producing this type of policy, the following are some notes should be considered. The policy should:

- Apply to all levels of staff, as well as contractors, consultants and other individuals the organisation might employ on a short-term or temporary basis;
- Provide a clear definition of what is considered personal use, including information on how much access time is acceptable and when personal access is allowed, as well as expressly stating that the user should exercise good judgment regarding the reasonableness of personal use;
- Provide a clear warning of what is strictly forbidden and prohibited, including the following:
    - Accessing pornographic or indecent websites and downloading material from these web sites;
    - Using chat rooms or online messaging services for the exchange of information, comments or opinions that are sexist, racist or any that will cause offence, and this includes which the use of offensive and indecent language;
    - Any activities or transactions that are fraudulent or meant to deceive, or where the user masquerades as another user;
    - Any activities or behaviour that might provide unauthorised access to other organisations' information, their systems and the network services they use, as well as causing disruption or denial of services or damage to or loss of information or processing resources.
- Provide warnings regarding the need to comply with any copyright, digital rights and licensing requirements and restrictions related to material on web sites;
- Provide instructions regarding the downloading of material from the Internet, including the cautionary measures to be taken to avoid downloading malicious code such as viruses, worms and Trojan horses and similar executable programs;
- Provide a clear statement of what disciplinary actions are to be taken if staff do not comply with the policy;

- State what the organisation's rules regarding its right to audit and monitor networks and systems on a regular or periodic basis. Staff should be advised that their access to the Internet and use of email may be monitored in accordance with the laws in whatever jurisdiction the organisation operates.

A note of what this might include might be useful to staff as a cautionary measure. For example, the fact that websites that have been used can be traced, the frequency of times they were accessed and time spent at such sites will appear in an audit trail, as well as the situation regarding using the Internet resources for personal transactions (e.g., online banking, bookings and orders). Consultation with legal and human resource personnel should be engaged in the production of these types of policies.

6.4.2.3 Specific Policy Provisions

These might include the following:

- Warning users not to reveal their account password to others or allowing use of your account by others, which includes those outside of the organisation such as family or friends in the case of a laptop being used offsite;
- Warning users to make statements regarding warranties, licences or any other statements that might legally commit the organisation, expressly or implied, unless it is a part of the user's normal job duties and prior authorisation to carry out this action has been given;
- Reflecting the organisation's requirements for protecting the information that staff create on their systems and what is deemed to be the acceptable (and legal) use of encryption for information that is considered sensitive or especially vulnerable;
- Monitoring and/or intercepting a communication network or service for the purpose of intercepting data (unless, of course, this activity is a part of their job function and prior authorisation has been given and it is allowed in the jurisdictions covered by these activities). For example, the UK Regulatory Investigatory Powers Act (RIPA) law covers some such activities.

6.4.2.4 Travel Company Case Study

The following is a sample policy of that used by the Travel Company. Use of this policy will require an organisation to customize the contexts to match their own requirements and legal obligations and liabilities. Policies such as this one should always be checked by someone that has some knowledge of the law and contracts, especially since what is in this example may not match the business rules and requirements such as the use of emails for personal use.

6.4.3 Information Handling Policy and Procedures

6.4.3.1 Basic Idea

The aim of these procedures for information security is to ensure that all staff know what they should do to handle information in way that protects its confidentiality, integrity and availability—whether it is the processing, storage and archiving, distribution, copying or disposal of information.

6.4.3.2 Information Management

Ownership, Accountability and Third-Party Custodianship

Information that is generated and/or created using the organisation's resources (money, staff, ICT systems and other information, especially that owned by the organisation) is generally deemed to be the organisation's property, and the organisation becomes the owner. As the ultimate responsibility for all of the organisation's assets are the senior executives and directors and CEOs, they are accountable for this information.

Often responsibility for the day-to-day management of the information is delegated down to particular groups, departments, functional units and/or individuals within the organisation. Custodians within these groups are then given individual operational responsibilities of being the "owner/custodian," which means they need to manage these assets and to ensure they are afforded an appropriate level of protection. In the area of data protection with the roles of data controller and data user, both these roles are associated with different information responsibilities. It might also be the case that information is managed and processed by an outsourcing contract to a third party. In all cases of delegation, custodianship and outsourcing the ultimate accountability of these assets rests with the directors and executives, and it is their responsibility to ensure their safe keeping and the organisation's well being.

Use and Misuse of Information

The ISMS involves an ongoing process of practicing the right things to do for the business with due care and diligence to protect information from a range of threats and risks. This means that staff must use the organisation's

Travel Company Email Use Policy

1. Scope and Objectives
1.1 The purpose and objective of this policy is to prevent damaging the image of the Travel Company.

1.2 This policy covers appropriate use of any email sent from the Travel Company email address and applies to all employees, service providers and agents operating on behalf of the company.

2. Policy
Prohibited Use: The Travel Company email system shall not to be used for the creation or distribution of any disruptive or offensive messages, including offensive comments about someone's race, gender, religion, disabilities or age. Also prohibited is the use of emails of a sexual and pornographic nature, religious and/or political beliefs. Staff that receives any emails of this type from any other member of staff or from outside sources should report the matter to their manager immediately.

Personal Use: The Travel Company allows a reasonable amount its email resources to be used for personal use as long as such use does not interfere with the user's normal work tasks and the company's day-to-day operations and customer services. Any nonwork-related emails, and attachments, shall be saved in a separate folder from work-related email. However, the sending of chain letters or joke emails from the Travel Company email account is prohibited. Any mailings sent via a distribution list from the Travel Company shall be approved by the head of the Travel Company operations. These restrictions also apply to the forwarding of mail received by a Travel Company member of staff.

3. Monitoring Email Usage
The travel company reserves the right to monitor email system usage and to check anything that is stored, sent or received via the email system without prior notice. Therefore members of staff should not expect any privacy protection from the company's email system.

4. Disciplinary Action

Any member of staff violating this policy may be subject to disciplinary action. Depending on the severity of the resulting security policy violation the action may include the termination of employment.

Employee name ______________________

Signature______________________________ Date _____________

Line manager
Signed ______________________________ Date _____________

Email Usage Policy Ref; TCP 3 version 3.0 (March 3, 2015)

information and resources in an acceptable and proper way to avoid security comprises and/or legal complications and actions (see Section 6.4.2).

Information Classification

Information should be graded appropriately according to its level of sensitivity and/or criticality. This enables an organisation to be able to deploy the right amount of protection for the information commensurate with these sensitivity levels. There are many names and levels of grading that can be assigned to information.

6.4.3.3 Information Gathering, Creation and Processing

Gathering and Collecting

Gathering and collecting information is a normal function of most organisations. This may be through market research, research on customer activities (types of spending, seasonal fluctuations), customer surveys and feedback and other forms of research. There are some legal, contractual and security implications about collecting information which organisations should be aware of and pay attention to. The most prominent one of course is personal data: what can be collected, what be done with this data and what controls are needed to protect it against abuse or misuse. Another area of attention is intellectual propriety and copyright related to both electronic and nonelectronic information and digital rights, digital rights management (DRM) and the rights of individuals.

Processing Efficacy, Effectiveness, Efficiency and Correctness

The activities of processing information can mean many things, but all have information security requirements to ensure the confidentiality, integrity and availability of information is preserved. This means doing the "right" things at the "right" time (efficacy of processing), getting these things done (efficiency of processing) and doing these things "right" (effectiveness of processing).

Most information that an organisation processes is vital to its business, whether it is processing customer orders, accounts and statements, online transactions (buying and selling products), formal company accounts and tax records or other business processes. It is important to get it right to avoid losing money, time, productivity, credibility and possibly customers through incorrect data entry and processing of this data and subsequent results with mistakes and errors. Some of these problems are caused by human error, lack of training in using processing applications, software flaws, system malfunctions and failures, data corruption and data loss or unauthorised modification.

It is estimated that a high proportion of business losses and security compromises result from the people who handle information. The correct processing of information is a major area of security problems resulting in customer dissatisfaction, business losses, inefficient and ineffective business operations as well as legal action for failure to honour contracts or to be compliant with legislation.

Offsite Processing

Some information and data might be processed offsite, such as a member of staff using a laptop to process information whilst out on business at a client's office, at home or in a public place. A managed services company, under an outsourcing contract, might process the information. Whatever the business case, the protection of this information during processing should be afforded at least the same level of protection as it would receive onsite.

6.4.3.4 Information Storage

Electronic

With today's technology, organisations and users have many ways of storing information:

- Network servers;
- Hard drives (internal and external);
- Optical disks (e.g., CDs and DVDs);
- Flash memory devices (e.g., USB sticks, SD cards, memory cards), CompactFlash cards;
- Tablets, iPads, iPods and MP3 players, and other similar technologies.

Whatever means and methods of storage are used, security should be addressed: ensuring that confidential information is protected to prevent unauthorized disclosure, that the integrity of the information remains intact (i.e., does not get changed, modified or flawed whilst in storage) and that the information is available to all who need it. A clear policy needs to be established on what protection measures need to be in place to protect its confidentiality, integrity and availability. The more sensitive and/or critical information should be stored using memory segmentation or paging, cryptographic technology and/or a strong access control system.

Paper Based

Information in paper-based form requires just as much attention as information in electronic form. This should include:

- Physically protecting sensitive and/or critical information by locking papers, documents and reports in lockable filing cabinets, desk drawers or in safes to reduce the risk to an appropriate level, as well as providing standard physical security to building, offices and rooms;
- Physical separation of papers having different information classification;
- A clear desk policy should be established to ensure papers and documents are put away when unattended;
- Papers and documents should be protected from physical or environmental threats and damage (e.g.,fire, water).
- Papers and documents kept offsite should have the same level of physical protection.

Onsite and/or Offsite

The decision of whether information should be accessed and stored onsite or offsite depends on the circumstances of the organisation. Sometimes offsite storage is a better option, as it may afford a better level of risk management. Management must take into account the cost of such storage and the availability and convenience of accessibility to such storage. Offsite storage should always have the same level of protection as that given to onsite storage.

Information might be looked after by a managed services company that stores and processes information on behalf of the organisation. Access to this information is facilitated by use of networking.

6.4.3.5 Information Sharing, Exchange and Distribution

Electronic

With today's technology, organisations and users have many ways of distributing information. Users can push information out (such as podcasting, broadcasting or server-push) to recipients or have it available on a network to enable recipients to pull the information down from a website or server. Whatever means and methods are used, security should be addressed so that:

- Distribution of sensitive and/or critical information is kept to a minimum, and the information should be clearly marked to indicate its classification;
- Confidential information is protected to prevent unauthorised disclosure or interception;
- The integrity of the information is prioritised to ensure the information does not get changed, modified or flawed whilst on route;
- The information is available to all those that need it;
- A clear policy is established on what types of information can be sent electronically (e.g., via email and email attachments) and what protection measures need to be in place to protect the more sensitive and/or critical information;
- If confidential information can be sent by those with the right level of authorisation, then cryptographic technology is used;
- The use of anonymous re-mailers and other technologies should be forbidden.

Fax

Today fax machines are very much less in use than they were 10 years ago; however, for those organisations still using such technology, care needs to be taken since sending information by fax machines can sometimes cause security problems.

- Make sure steps are taken to ensure faxes are indeed delivered to the right destination. It is common for a sender to type in the wrong telephone number or hit the redial button, and the fax arrives at the wrong place.
- If sending sensitive business information, also make sure the fax machine is physically secure, that an authorized person is attending the machine as the fax comes through and have the person telephone or email to confirm that the fax has been securely received and the correct number of pages has come through.

Post and Courier Services

When sending information by post or by courier services, there needs be a clear policy stating what types of sensitive and/or critical information can be sent and the means and methods for doing this:

- Information at the highest level of sensitivity should use the technique of double enveloping, with the outer envelope not indicating the classification of the contents but the inner enveloped marked;
- It is generally accepted practice that envelopes containing restricted information could be sent in an single envelope as long as it doesn't indicate the classification marking but is marked with the words "to be opened by the addressee only";
- The use of tamper-proof containers or boxes, or such containers that make it obvious that an attempt at access has been made;
- Using registered or recorded mail services;
- Using trustworthy and accredited couriers.

6.4.3.6 Information Destruction

Erasure and Deletion of Information from IT Systems

Once information has gone past its "sell by date" and no longer needs to be kept, it should be deleted and/or destroyed from the system.

- Electronic files and folders on PCs, laptops and servers should be deleted. This can be done using the standard operating system commands or application functions or by using an appropriate piece of software technology that wipes clean the information.
- A word of caution: although most systems have a delete command, the information is not actually deleted—only the link to the directory system is actually deleted or disconnected, which means it still is possible to recover part or all of the information if you know how to locate in the system. There are ways of improving this situation other than by physical destruction of the storage media.
- Deletion of information should also apply to all copies as well as backups of the information.
- Having an orderly file directory system helps to facilitate deletion of files and any copies or backups.

- The asset inventory should be revised to record the deletion of information.

Disposal of Hardware and Storage Media

Disposal of IT systems poses the problem of information contained in their memory systems.

Disposal of Paper-Based Information

Again, as with electronic information, once paper documents and files have gone past there "sell by date" and no longer needs to be kept, they should be suitably disposed of.

- Paper documents, reports and other paper-based information should be appropriately destroyed using approved shredding or other physical destruction methods and technology (e.g., burning, pulping).
- Destruction of paper documents should be carried out by a trusted person or an approved external organisation.
- Copies and backups of all information being disposed of should also be destroyed.

6.4.3.7 Information Backups

Identification of Information to Be Backed Up

All information should be backed up one way or another. Care should be taken to back up information on staff PCs and laptops, information on servers and on other IT equipment.

Methods of Backups

How, when and where backups need to be done must be defined. Backups can be done centrally using a network of one or more servers, servers at a specific department level or locally at a user's computer. The policy should define which information applies to which type of backup. For example, operational information is commonly done via a central network server, whereas locally processed information might be done on someone's PC or external hard drive or memory device. It must not be forgotten that it is not only electronic information needs to be backed up, but also paper-based information systems since all information systems, IT or non-IT, suffer from the risk of physical or environment damage or loss as well as theft.

How frequently backups should be made should be related to the criticality value or level of the information. Daily overnight backups are quite common in many organisations, especially the large ones. End-of-week backups should be the norm and may be combined with the daily overnight process. Backups need to be done often and regularly, and the more often the better, within reasons of practicality. The more time information is left not backed up, the more risk the organisations run of it being lost or damaged.

Protection of Backups

Critical and/or sensitive information should be:

- Secured to protect its confidentiality and integrity with the use of cryptographic technology or strong access controls;
- Physically protected to ensure the integrity and availability of the information.

Testing Restored Backups

There should be a procedure to make sure the backup process and system is working properly and the backed up information can be read and is a faithful copy of the original. A good maxim to remember is that your backed up information is only as good as your ability to restore/recover the information from its backed up state without it being corrupted or lost in any way. What follows from this is that the system and media you are using for backups must be of a good quality and the process of restoration needs to work properly. If you can't restore your valuable business or customer information to its original state, then it was not worth the time to back it up in the first place. Testing the backup system should be done if there are any changes to the backup process and/or business environment using this process. Testing the process should be undertaken on a regular basis, regardless of whether or not there have been any changes. It is better to find out how good your backup and restoration process is before a major disaster or incident happens than find out later that it does not work properly—always be ready and prepared before an incident not after.

Storage

Storage of the backups is another issue that needs to be addressed by the policy. The normal best practice for this is to store the backups offsite somewhere away from where server and processing of the information takes place. The reasons should be quite clear: if such backups are kept on the same site, then they face the same risks the originals are subject to (e.g., damage and loss by office fires, physical destruction, theft and other threats). Hence, backup means not only back up the information but also back up the information somewhere else.

Store some backups away from home (or offsite) in case your backup data gets stolen or damaged along with the computer it is backing up.

Storage Media

The type of media used for backups is also another policy issue. There are many types of media for this purpose, including tapes, DVDs/CDs, USB sticks, SD cards and other removable memory devices as well as removable or external hard drives.

Backup Records

Accurate records should be kept of what is backed up, where the backups are kept, date/times and other information that might be necessary for incidents, disasters and system failures, as well as for auditing purposes and other information-handing processes.

6.4.3.8 Information Records Management

Records

Keeping records is an important feature of management systems, including ISO/IEC 27001. Records provide a vital link between past events and activities, and the effectiveness and efficiency of the ISMS to preserve information security and to manage the organisation's risks.

Retaining or Archiving Information

Certain documents an organisation has will need to be kept for a certain period of time as required by law, such as tax records and company accounts. For documents outside of the legal and regulatory categories, an organisation may have both business and technical reasons for the retention of information, such as:

- Protecting its IPR, trade secrets, and research results;
- Audit trails, system and network configuration files and other records, especially if they are required to be used as part of an incident handling or a forensic and/or criminal investigation.

6.4.4 Access Control Policy, Procedures and Processes

6.4.4.1 Basic Idea

Access control is a very broad subject and encompasses various types of methods and means of access such as having:

- The rights and means of access to enter buildings, offices and computer rooms;
- The right, privilege or opportunity of access to use or deploy something like an IT system;
- The right, means and ability of access to approach, see or talk with someone such as might the case with using the organisation's IT resources including laptops for online telephone calls and video messaging;
- The right and means of access to act on and/or process information (i.e., obtain, retrieve, copy, store or destroy to prevent unauthorized access or accidental damage to or compromise of information);
- The ability to be reached, such as having access via a mobile computing device.

Access is given to information and organisational resources to enable staff to carry out their particular job functions. Hence, access should not be given to specific information if it is not necessary for the conduct of the staff's official business duties. Normally the "principle of least privilege" is used to grant users/staff permissions they need to do their jobs, thus not providing additional permissions that are not necessary to minimise the risk of compromise. This principle should not be mistaken for the other principle of segregation or separation of duties, even though this later principle employs the idea of giving limited or least privileges.

Of course, this does not mean compromises will not occur with those users that have been given access permissions and rights, as there is always the case of those authorised to have access doing damage to the system and its contents. This can be the case with insider trading, staff involved in internal fraud, theft by staff and other incidents where the culprits did have authorisation. Hence ISMS measures need to protect against those not having sufficient access rights and permissions and those owners of the information to control those that do.

Although access control is primarily about ingress—the rights, permissions and privileges to gain entry—there is also a dual protection aspect, and that is the egress situation where information exiting from a building or system does not leave in a condition that could be compromised.

It is important from the word go to have a clear understanding of the business requirements for access: access to what and by whom and the means of gaining access. What are the requirements for access to the different information systems the organisation has? Are there requirements for access to different applications and processes? What are the physical access requirements for staff, contractors, maintenance engineers or visitors? How is access control to be implemented regarding IT-based information systems? At the technical/IT implementation level, access falls under particular policies—mandatory access control (MAC) and discretionary access control (DAC). MAC policy is centrally controlled by a system administrator, whereas DAC policy is generally controlled by users.

6.4.4.2 Access Rights and Information Classifications

The organisation needs to decide what access rights are given to information and by whom. Some may be given access to information at different levels of classification. Staff will be given access to information that is on a "need-to-know" basis.

The higher the classification (most sensitive information), the smaller the group that has need-to-know rights; hence, the biggest group of users would be those having access to "restricted" information (less sensitive) (Table 6.1).

In addition, information can further classified (e.g., a confidential marking might have a caveat attached to it such as in Table 6.2).

In addition to need-to-know rights, which are a means of allow users to read information, there are also need-to-modify rights, which give users permission to modify, delete and/or generally create information.

An example of these different types of access is that referred to as read and write access. There are, of course, other access rights, such as the right to be able to execute a particular piece of software code. Table 6.3 shows

Table 6.1
Example Classification Rights

Marking	Rights
Strictest in Confidence	Information available on a strict need-to-know basis (e.g., available only to the CEO and the board of the company)
Confidential	Information available on a need-to-know basis (e.g., available only to senior management, or available to a group in the organisation with a specific job function such as finance, human resources, research and development)
Restricted	Information that is not available to the public but is generally available internally to all employees or staff
Public	Information available to the public (e.g., information about a company on its website)

Table 6.2
Example Classification Caveats

Marking	Rights
In Confidence	Need-to-know rights covering information without a caveat
Medical in Confidence	Need-to-know rights with the caveat that medical information is restricted to those working in the health care field—doctors and nurses
Personnel in Confidence	Need-to-know rights with the caveat that personnel information is restricted to those working in human resources
Financial in Confidence	Need-to-know rights with the caveat that financial information is normally restricted to the finance department/group, the CEO and directors

Table 6.3
Example Access Rights

Access Right	Access: Subject → Object
Read (R) access	The subject can Read the contents of a file List directory contents
Write (W) access	The subject can change the contents of the information contained in a file or directory with these tasks: Add Create Delete Rename
Print (P) access	The subject can print a copy of the file.
Copy (C) access	The subject can make a copy of the file.
Execute (X) access	If the file is any software program or executable code, the subject can cause the program to be installed and run.

a list of these rights found in applications and form part of MAC and DAC policies. Some applications do not offer all these options or restrictions.

6.4.4.3 Authorisation, Accountability and Ownership

At a management system level, it is normal policy and good security practice for information (system or process) to have an owner. This owner is responsible for making sure that the information they are responsible for as owner is adequately and suitably protected. Thus, the owner is accountable for ensuring they have appropriately classified the information, they have compiled an information asset inventory and they will ensure the necessary and appropriate control measures and actions are taken to protect the information at the defined information classification. They need to ensure that such protection is implemented in order to manage the risks to this information. The technical implementation of access control is, of course, outside of the scope of ISO/IEC 27001. Authorisation for access is also another management aspect of ownership of information: authorisation granting or denying access and the appropriate access rights.

The information owner might be, for example, an individual who is head of a group, unit, department or division of an organisation.

6.4.4.4 User Access Registration and Deregistration Process

The implementation of access control does have many variants depending on the type of access, the methods of access and the technology being used. There are, however, some common elements that should be considered when producing access control procedures.

The overall process should start by registering the user onto the system and then setting up an account from it. Once this has been established, the user can use the account to access resources on the system according to the rights and permissions they have been given. This also applies to the access of services (e.g., telecommunication services and Internet services).

The management and maintenance of access policy and procedures; the rights, permissions and privileges; the control measures and methods for access—these are not necessarily trivial and can sometimes prove to be quite an administrative burden with large overhead, if not managed correctly, especially in large organisation with thousands of users and system accounts.

Systematic, properly implemented ISMS processes and the effective deployment of ISMS measures can reduce this workload. From the point when a user is first registered and given an account on the system until the time when that account is cancelled and the user's access rights need to be revoked, there needs to be management measures in place to ensure the access process is secure, effective and robust. This process involves a number of measures, both nontechnical and technical, all of which should work in harmony and without compromise.

6.4.4.5 User Access Termination Process

Once employment has finished or has been terminated, or there has been any change in employment, a number of internal security-related activities need to take place and be enforced regarding user access (see also Section 6.4.5). This applies to the termination of employment of full-time staff as well as contractors. Such activities include, but are not limited to:

- Return of organisation assets (e.g., IT equipment, mobile phones and other devices belong to the organisation; information; documents, files);

- Removal of access rights and privileges;
- Enforcement of ongoing legal responsibilities contained in confidentiality agreements and terms and conditions of employment continuing for a defined period after termination.

Requirements regarding the termination of employment related to security aspects should be addressed in the contract of employment. Similarly with third-party contractors and agency staff, security requirements regarding termination should be included in their contracts.

6.4.5 Human Resource Policies, Procedures and Processes

6.4.5.1 Basic Idea

An organisation's human resources (or personnel) group is generally there to recruit, develop and utilizes its personnel for the purposes of meeting the organisation's operational and business needs.

It has been shown many times through work experiences that the biggest threat to information systems originates from people. The people threat could be accidental (e.g., errors in entering customer information into an ordering system or the unintentional deletion of files on a PC) or intentional (e.g., the abuse of organisational ICT systems for personal gain or destruction of critical data as might be the case with a disgruntled employee).

There are three important stages to consider in the life-cycle of employment: (i) recruitment (Table 6.4), (ii) during employment (Table 6.5), and (iii) termination of employment (Table 6.6). Therefore, management should focus its efforts on getting the people aspect of information security right and management commitment and leadership to properly and appropriately address these three stages.

6.4.5.2 Training and Awareness

Information security training must achieve not only knowledge transfer and understanding security policies and procedures, but should also aim at getting staff involvement and commitment to adopting the procedures. But just knowing the procedures without practice does not modify the risk of loss or damage to the organisation's information. Some specific information security training and awareness aspects (the content of which should be as a minimum: the information security and business risks, rules and guidelines, roles and responsibilities and commercial and legal aspects), include:

Table 6.4
Recruitment Activities

Activity	Commentary
Recruitment	The human resources group is responsible for facilitating the recruitment of the right people to fit job vacancies. Defining employee skills, qualities and competence is an organisationally determined aspect. In the area of information security, there are some specialist jobs such as a security officer/ manager, risk manager and IT security expert. In addition, there will be staff that are not specialists but will need to know about, and how to use, information security policies and procedures in their day-to-day work.
Screening	This process involves assessing the applicant's suitability for the job advertised. This normally involves screening applications based on their work experience, skills and training for the described job. This normally involves a two-/three-stage process. Screening comes first at a general level (e.g., reviewing applications, requesting a CV), then may be followed at a secondary level by getting letters of reference from previous employers and from educational establishments or certificates as proof of their qualifications). Depending on the type of job being applied for, further screening stage might need to be carried out. For example, for some security-specific jobs may require more detailed assessment of their previous work and/or social background. This could go as far as checking whether the individual has a criminal record or checking their financial state by doing a credit reference check.
Interviews	Like most recruitment drives, the interview activity is a critical part of the process involving the human resources group, representatives from the department involved in the vacancies and the candidates themselves. The more detailed screening process normally comes after the interview relating only to those that have been sort listed.
Signing documents	The new employee will be expected to sign a contract of employment with the employer's terms and conditions. They will also be asked to sign a nondisclosure or confidentiality agreement, as well as various forms for their salary payments and whatever security forms are deemed necessary (e.g., an acceptable use policy).

- Processing of personal data and/or the organisation's confidential and critical information;
- Information processing activities;
- Use of IT systems and software, the Internet and email, mobile and WiFi services;
- Reporting incidents, system malfunctions and failures;
- Backup of data;
- Physical protection;
- Dangers of social engineering.

Table 6.5
During Employment Activities

Activity	Commentary
Employee responsibilities	Staff will be assigned certain roles, responsibilities and duties during their employment, which are related to the organisation's information security policies and procedures. These include general security duties and responsibilities as well as more specific security duties and responsibilities.
Employee system account	During employment users would normally be given a system account to allow them to access the organisation's IT systems and networks. What they can access and what rights they have to read, write, print, copy or execute files on the system will largely depend on their specific job role. It is good practice for the human resource group to keep a record of this on their system.
Provision of training	Staff needs to be given training according to their job functions and needs. The objective of this should be to provide: Awareness of information security and risks; Instruction on how to use security-related policies and procedures; Skills for career development. Training can take various forms, such as on-the-job training, formal training courses, online training and at the desk self-instruction using training software.
Review of training	As business needs change, staff development and/or staff get moved to different job functions and/or responsibilities and training needs will likely change. This will entail reviewing staff training, providing additional or new training and updating the staff records. It is important that staff keep up to date with the organisation's developments in information security.
Disciplinary process	Security compromises involving staff can require disciplinary action depending on the severity of the incident.
Legal representation	Some information security incidents can involve civil or criminal activities. Members of staff who are named as being the perpetrators or being involved in any way could face legal action against them as individuals and possible action against the organisation itself.

6.5 Sector-Specific Controls

A number of sector specific standards have been developed as part of the ISO/IEC 27001 family of standards that specify controls and implementation guidance for individual sectors. These include:

- ITU-T X.1051 | ISO/IEC 27011—information security management guidelines for telecommunications organisations based on ISO/IEC 27002 ISO/IEC 27015—information security management guidelines for financial services;
- ITU-T X.1631 | ISO/IEC 27017—guidelines on information security controls for the use of cloud computing services based on ISO/IEC 27002;

Table 6.6
Termination of Employment Activities

Activity	Commentary
Termination responsibilities	Staff employment gets terminated for various reasons: (a) voluntary termination (e.g., if the member of staff wants to move or they need to move for domestic reasons) or (b) involuntary termination such as being fired for disciplinary reasons or being laid off. For these different types of circumstances, the responsibility of the handling of the termination of staff employment should be assigned to someone in the organisation's human resources group with the appropriate seniority level. This role should involve making sure the process is handled correctly to ensure that all business, legal and personal issues are concluded in a proper and effective way appropriate to the circumstances of the termination. Some terminations can be the result of disciplinary action and a severely disgruntled employee with bitter feelings toward the organisation. This type of situations need to be handled sensitivity and with extreme caution, with cognizance of what damage the employee might cause before leaving. In some organisations, employees are asked to the leave the premises immediately without serving the normal period of notice if there is a risk of a security compromise.
Return of assets	Staff is normally expected to return all assets that belong to the organisation. This should especially be the case with all computing equipment, storage devices, software and applications as well a paper documents, files, reports, manuals and/or procedures containing the organisation's information, in particular that which is sensitive, confidential, restricted or related to personnel.
Removal of access	To avoid the risk of unauthorized access to buildings, offices and rooms, pass cards, identity cards and other devices should be collected from the member of staff. Also any physical entry PIN codes should be removed from the system. All access accounts to information systems, in the name of the member of staff, should be removed from the system. This is to avoid unauthorized access by the ex-member of staff or a colleague or accomplice at a later date. This applies to all ICT equipment from system servers, PCs, laptops, mobile service accounts and any other devices that allow a user to gain access to the organisation's information.

- ISO/IEC 27018—code of practice for PII protection in public clouds acting as PII processors;
- ISO/IEC 27019—information security management guidelines based on ISO/IEC 27002 for process control systems specific to the energy utility industry.

These standards provide sector organisations with a source of controls they can use during the risk treatment phase; that is, when determining a set of controls that will help in modifying the identified risks (ISO/IEC 27001 Clause 6.1.3 b)).

6.6 Benchmarking with ISO/IEC 27001:2013 Annex A

The set of controls that has been determined according to ISO/IEC 27001 Clause 6.1.3 b are then checked against the list of controls given in Annex A of ISO/IEC 27001 to see whether any control areas or controls have been overlooked. This is merely a benchmarking exercise and not a replacement for the determination of controls exercise.

Even if there are controls in Annex A that have not been determined as necessary, this does not mean that they need to be added to the list of controls to be implemented. On the other hand, there may be controls that have been overlooked, and consideration should be given as to whether they should be implemented. In all cases any controls that are not included in the implementation list but appear in Annex A do need some justification as to why they are not included.

CHAPTER

7

Contents

ISMS Operations

7.1 Operational ISMS Procedures

7.1.1 General

As indicated in Chapter 6, there are several types of procedures that an organisation might need to have in place to protect its day-to-day operations. The operational procedures used by the organisation will be unique to it and its business operations. Operational procedures are generally detailed, written instructions to achieve a consistent level of information security performance of a specific function, process or operation within the ISMS scope. They are deployed to assist and ensure a level of information security commensurate with the level of identified risk.

7.1.2 Example Procedures

There may be operational procedures that are standard across the organisation, and some procedures may be specific to certain departments (e.g., call-centre procedures detailing the information security controls to be deployed by call-centre staff). The following are some of the procedures that an organisation needs to consider for implementation:

- Acceptable use policy and procedures;
- Access control policy and procedures;
- Backup procedures;
- Business continuity procedures;
- Call-centre procedures;
- Capacity planning procedures;
- Change control procedures;
- Communications and reporting procedures;
- Information handling procedures;
- Information security incident procedures;
- Malware procedures;
- Monitoring procedures;
- Recruitment procedures;
- PII procedures;
- Procedures related to secure operational areas;
- Service management procedures;
- Software maintenance procedures.

7.1.3 Training, Awareness and Usage

It is essential that the organisation's management not only makes employees aware of the procedures that are applicable to the job role and function, but also arranges for employees to have some form of training in the use of these procedures. A new recruit could easily compromise the information security of the organisation by making errors or doing the wrong things by lack of understanding and knowledge about the correct security procedures to follow in their day-to-day job. This may not just happen to new recruits, but also employees that have been in the organisation for a while may need to use a new procedure or a revised version of an existing procedure. In all cases the same applies, a lack of understanding and know-how can lead to an incident and hence compromise information security of the organisation.

It is also important that all employees use the procedures provided by the organisation to ensure consistency and the intended performance. If employees use their own methods of doing things in an operational environment, then this could not only compromise the information security of the organisation and its systems, processes and services, it would be against the organisation's policy and rules and could be rewarded with disciplinary action. Finally employees using nonstandard procedures are likely to result in a nonconformity being given in an audit. We should remember that a well-written procedure would be generally designed to achieve the most effective level of information security, and so the instructions should be carefully thought out and written in a way that will achieve this level. So a nonstandard procedure used by some employee may not only fail to achieve this level, but would be against the organisation's policy and wishes.

7.2 Ongoing Risk Management

A risk assessment is always just a snapshot in time based on the perceived risks and impacts at the time the assessment was carried out. There should be no doubt that business conditions will change, and with this the risks and impacts related to the organisation will change. So it is imperative that the organisation keeps up to date with its risk profile by reassessing its risks and impacts on a regular basis. What is regular depends on several factors including:

- Whether the organisation is operating in a dynamic environment, in which case it is most likely subject to frequent changes in business conditions—vitality of the market, its growth rate, its diversity and other variable conditions;
- The complexity of the organisation—in terms of processing resources; the volume, type and nature of information involved; the number interfaces and dependencies it needs to manage and take account of and so on; these are some of the risk factors and risks it needs to consider;
- If the nature of the organisation's business places it in the front line for attack, then the frequency of attack would be higher.

There is no set standard time for reassessing its risks and impacts. Some organisations do a reassessment every three to six months; some leave it to nine months—the actual frequency can only be determined by the organisation itself. It should be management that decides the frequency. In addition to a regular programme of risk assessments, then there will always be the chance that a risk-assessment needs to be carried out sooner than the next scheduled risk assessment. This might be after a major information security incident, a major system disaster, a major change in the organisation circumstances, or a management review (see Chapter 8) or ISMS audit (either an internal or an external certification audit—see Chapter 10). An organisation should implement a change management process (see Section 7.4.4) that includes information security aspects.

7.3 Operational Threats

7.3.1 Malware

7.3.1.1 General

Malware attacks are worldwide headline catching news. The virus is an example of a piece of malware or malicious software that is designed to infiltrate or damage a computer system. This class of software also includes worms, Trojan horses, spyware and adware, to name but a few pieces of malicious software code.

Malicious code is one case of those pieces of software that is designed specifically to do damage and harm, and put organisations infected with such code at risk. As is well known by many cases, viruses once in the system attach themselves to applications software, then replicate and spread themselves very fast like a fire in a forest that has not seen rain for weeks or months. In their path through the system, they cause destruction and damage just like the fire. Even when they have been discovered. they may not be immediately eradicated—just like a fire that taken hold of the forest, extinguishing the flames might take sometime. So for a period of time, the virus is out of control. The virus might be deleting files, damaging software programs and applications and may even be aimed at reformatting the hard disk. Even the more benign types of viruses can cause the organisation problems since continual replication of the virus can eat up memory, trigger erratic behaviour and maybe make the system crash—all resulting in degradation of performance and possible a denial of service.

The worm is a self-replicating piece of software code that does not attach itself to an application, but nevertheless causes widespread damage across networks such as the Internet. Since it is self-replicating it eats up network bandwidth, consuming computer system memory, slowing down systems and in the worst case causing a widespread denial of service attack. Worms such as the "Morris Worm" of 1988 cause global disruption and chaos to many systems connected to the Internet at the time the worm invades these systems. Several of the worms that are released onto the Internet only replicate and spread themselves; however, some worms have been produced to carry "payloads," such as to delete files, send files via email to other systems, install "back doors" in systems and even encrypt files. The back-door problem can result in an opportunity for the attacker to create a "zombie computer" within the infected system. When similar zombie computers are networked together to form a "botnet," which subsequently can be used by those responsible for sending spam, they can create several security incidents.

The Trojan horse is a piece of software code that contains and/or installs a piece of malicious code into a system. This piece of malicious code is often called a payload or simply a Trojan. These Trojan horses might appear as useful and harmless pieces of software, but beneath the surface lurks the malicious code that, when the seemingly useful piece of software is executed, triggers a sequence of damaging events. The Trojan horse can cause a lot of unauthorised activities to occur on the victims' systems without their knowledge or consent such as:

- Downloading and uploading files from the Internet;
- Sending out emails using the victim's email address book, including the distribution of spam;
- Phishing in the victim's files to gain access to credit and banking details;
- Causing the system to restart;
- Attaching the system to a botnet;
- Deleting and corrupting files.

Trojan horses can infect systems in several ways such as via an email attachment or by a user downloading a file from a rogue site on the Internet.

7.3.1.2 Managing the Problem

Protection from malware needs proper management processes and controls in place. For a start, the organisation should produce a policy prohibiting the use of unauthorised software and then implement controls to prevent or detect the use of such software. The organisation should also have a policy regarding the downloading or receiving of files and software.

A set of procedures should be produced by the organisation for how to deal with and report malware problems. Responsibilities for dealing with malware should also be defined, and this should be included in these procedures. Staff should be trained in how to deal with malware.

The organisation should conduct regular audits on the software that supports their critical systems and processes. Such exercises should search for the existence of any unapproved files or software.

The organisation should implement software that is capable of detecting malware on computer systems and media. This type of software should be capable of scanning files, email attachments or web pages for the presence of malware infections. There are a number of commercial products on the market for this purpose. The functional capability of these products does vary: generally they are designed to detect, prevent and eradicate malware from computer systems and other IT devices. They generally scan the systems files and folders located on hard drives and other memory devices. The malware software relies on a *dictionary* of known pieces of malware to compare against when scanning the system. The effectiveness of this software relies on keeping the *dictionary* up to date: most product vendors have an online facility to provide live updates. Of course, this live update needs to be done on a regular basis—at least once a week, if not several times a week. It is a good idea to set your system to do an automatic update.

As a word of caution, the malware code writers always try to stay one step ahead of the vendors by using various techniques to disguise the malware to circumvent the dictionary match process.

Of course payloads such as those attached to Trojan horses are more difficult to protect against. Most of these are delivered by email, and there are some email best practice measures that could help:

- Being alert to emails that are received from someone that the user does not know or that has suspicious or unknown attachments;
- Set the email preferences to not automatically open attachments;
- Update the antivirus software automatically;
- Update the operating system with the latest patches.

7.3.2 Unauthorised Access

7.3.2.1 General

Unauthorised access to information can lead to disclosure, modification or unauthorised use of the information. Such access might happen as a result of an internal compromise of organisational policy and abuse of privilege, or it might occur as a result of someone externally breaking through organisations security systems. Unauthorised access to information might occur by breaking through layers of physical protection or electronic protection, or a combination of both. Of course, access via the Internet or other network is a common problem. Whatever the means and method of access, once the unauthorised person has gained access to the organisation's information, they have an opportunity to steal, modify or destroy this information.

Depending on the type of information being accessed (sensitive, critical or personal), then the organisation may be facing a range of problems. For example, if the unauthorised person has gained access to customer details including passwords, PINs, or bank account details and sells or publishes these details online, then the organisation faces commercial and legal problems. Another example might be the unauthorised person that gains access to critical information that controls an industrial process and the same person modifies the information or takes control of the process—this could be a serious safety problem as well as potential commercial and legal problem.

An unauthorised access attempt might be from an outsider (e.g,. a hacker, competitor, industrial spy) trying to break into the organisation's system and gain access or it might be an attempt by an insider abusing their privileges or using their inside knowledge of the organisation to gain access to areas of the organisation they have not been given access to.

7.3.2.2 Managing the Problem

The problem of unauthorised access to information can be managed at different levels. The organisation should have an access control policy, and it should have specific procedures for different access control methods and for covering different types of access (e.g.,access to information, applications, processes, IT and networks). The control methods implemented to manage access should be strong enough to prevent unauthorised access. Control of these methods and the information used for gaining access needs proper management and administration. Controls need to be in place relating to who has access, for what purpose and to what information is access given, via what devices can access be made: establishing authorized computer, system and network accounts, password and privilege management, access restrictions, reviews of access rights, disciplinary action regarding abuse of rights and training of users on access control methods, rights and privileges.

7.3.3 Insider Threat

7.3.3.1 General

The insider threat is a malicious threat to an organisation, which is becoming more and more a common occurrence. Such a threat could be from current staff and employees from within the organisation, but it might also be from ex-employees or even contractors or business associates that are working closely with the organisation and have certain internal access privileges because of the nature of the work and collaborative ventures they are involved in.

The insiders, employees or members of staff will have computer accounts given to them for legitimate, work-related purposes to undertake their job function and work-related duties. This means they already have some level of authorised access to computer systems to carry out certain job-related tasks, unlike the outsider who does not have such access (the case of the onsite contractor is different, and they may be given some access rights). The insider may abuse or misuse these privileges and cause harm or damage to the organisation. In addition, the insider generally has far more knowledge, experience and familiarity with the workings of the organisation: where things are, who does what, how things operate, the strengths and weaknesses of the organisation, the rules and procedures. With this inside knowledge of the organisation and the privileges they have, the insider has much more scope for attacking the organisation and being successful than the outsider, although we should not underestimate or ignore the power of the outside threat. Hence, it easier for the insider than the outsider to circumvent any information security the organisation has in place.

Some of the insider attacks are for personal gain; others are for revenge. Both of these cases are intentional and deliberate attacks. On the other hand, an insider threat might be due to user error or mistakes (i.e., accidental or unintentional).

A member of staff might be using the organisation's resources for their own personal gain: running their own private business for monetary profit, selling commercial information to the organisation's competitors, committing fraudulent transactions, stealing intellectual property, selling stocks and shares and so on.

A member of staff may be disgruntled and unhappy with their working conditions or feel that management or their peers have treated them unfairly. In this case, they might take out their dissatisfaction or anger regarding their situation by damaging system resources, deleting or corrupting information, leaking information, destroying files, deliberately upsetting customers, obstructing and disrupting normal business operations and other acts of sabotage against the organisation.

7.3.3.2 Managing the Problem

The insider problem can be managed at different levels, with a combination of the following:

- Policies and procedures—include instructions and information regarding insider threat;

- Training—include insider threat awareness into staff training;
- Recruitment process—check employment references, check for past criminal offences, review reports of suspicious or disruptive behaviour;
- Personnel management—manage conflicts; resolve disputes in the work environment; review working conditions of disgruntled staff; provide counseling and dispute resolution opportunities for disgruntled staff; be vigilant to unusual behavior and watch for unusual work patterns against what the organisation defines as standard, work-related activities and patterns of normal work operations;
- Ensure the risk assessment and risk treatment takes account of the insider threat problem;
- Termination of employment—implement a secure employee termination process;
- Implement controls for separation of duties and least privilege; enforce strict password practices; ensure secure backup and recovery processes; implement effective change control procedures;
- Access control—implement strict access controls; enforce strict access restrictions; ensure effective management of privileged users, including effective monitoring, logging and audit of system access; impose restrictions on remote access; review access rights;
- Online activities—be vigilant to online employees behaviour; manage usage of online services such as social networking through acceptable use policy;
- Enforce an appropriate policy and procedures for BYOD;
- System vulnerabilities—keep the door closed to unauthorized access and exploitation of system vulnerabilities by fixing, repairing and safeguarding against system vulnerabilities.

7.3.4 System Availability

7.3.4.1 General

If the organisation wants to keep its business running and to operate productively, then it needs to ensure the resources it needs to do so are available. This includes all types of resource: people, information, processes, services, networks, IT and facilities. A factor in this is how reliable, for example, is the organisation's network in providing for its operational needs? Does the network fail? Does it slow down? Does it provide for 24x7 uninterrupted usage? The backbone of most businesses is that of IT. The organisation might suffer from a poor record of IT system failures. Why? Is it lack of adequate maintenance, failure to install the latest software upgrades and patches, or is it that the demand for a specific IT system far outweighs its capacity to deliver against these demands?

If an organisation's IT system fails or its online presence is interrupted, it could suffer many hours of downtime, which translates into impacts such as loss of revenue, loss of customer confidence and possibly penalties due to a breach of contractual obligations if the loss of availability stops the organisation from being unable to honour such obligations.

Organisations can improve their chances of having the resources they need available as and when they need these resources. In particular, it is the resources that provide, processes and support its information, as well as the availability of its information that is the main issue.

In the case of IT and network resources, it is a case of regular maintenance and testing of these resources to lower the chances of IT system failures and outages, network service disruptions and protecting these resources from attack (e.g., a denial of service attack ,which can overload, slow down and generally disable systems and their ability to function normally). This is particularly the case with software maintenance, upgrades, patches and vulnerability testing, but can also apply to the hardware.

7.3.4.2 Managing the Problem

The availability problem can be managed at different levels including the following:

- Information—protecting information from unauthorised disclosure or modification, access to information, backups of information, information recovery in the event of a system failure, procedures for handling information, ownership of information assets and authorisation to access information;
- Services—management of service delivery, service provider contracts and SLAs, managing the connections to services, access and usage of services (e.g., networks, web services, managed security services, disaster recover services, outsourced services, teleworking services, media disposal) and authorisation to access services;
- IT system usage—effective access controls, restrictions of usage to avoid misuse, policy for acceptable use and procedures for correct use, secure processing of information using IT, employee terms and conditions on the use of IT, network services and the handling of information, authorisation to access IT systems;

- Software (S/W) management—restrictions on S/W installations and usage; control of S/W versions, upgrades and patches; separation of private and business use; control of S/W support and maintenance; use of default/secure configurations, access to utility versus application S/W; restrictions on remote use of S/W, control of S/W licences; separation of S/W in operational environments from S/W in development and testing environments; malware protection; S/W backups; secure control over operation S/W;
- Vulnerability management—risk assessment and impact analysis of the vulnerabilities, understanding and identifying system vulnerabilities, assessing and reviewing the vulnerabilities, taking correct or preventative actions to remove or diminish the weaknesses or to improve protection to avoid the chances of exploitation, improvements and controls to manage vulnerabilities need to be tested, ensuring S/W patches are from an authentic source, assuring the vulnerability process works in collaboration with the incident handling (Section 7.5) and business continuity processes (Section 7.6).

7.3.5 Social Engineering

7.3.5.1 General

Social engineering is a type of information security attack that exploits the psychological vulnerabilities of people by manipulating them to perform activities that result in divulging information—the "art of hacking into the human psyche." This type of attack can be performed online (e.g., emails, phishing attacks, social networking) or offline (e.g., over the phone to obtain information or posing as officials or service technicians to gain access to company premises, facilities and information). This attack is a confidence trick that aims to obtain information to gain system access to steal commercially sensitive information or to commit a fraud, a scam and so on.

7.3.5.2 Managing the Problem

The social engineering problem can be managed at different levels including:

- Consideration of social engineering attacks in the risk assessment and risk treatment exercises;
- Identification of information as sensitive, critical or personal, and evaluation of the exposure of this information to social engineering attacks;

- Establishment of a policy and procedures for handling social engineering attacks;
- Employee training to understand the problem of social engineering and to recognize situations that might be an indicator of, or precursor to, a social engineering attack and understand how to apply company policy and procedures to social engineering attacks for protection information;
- Occurance of regular audits and reviews of the controls in place to counter social engineering attacks—people, processes, procedures and access control methods;
- Establishment of roles and responsibilities for handling sensitive, critical or personal information;
- Keeping up to date with the latest information on social engineering attacks, feeding this information back into the training programme and using this information in the revision of policy and procedures.

7.4 Operational Processes

7.4.1 Protecting Information in the Operational Environment

It is important that the organisation has the necessary management controls in place to support and secure its operations and the information being processed during its operations. These controls include the following:

- Procedures for all operational activities associated with the processing of information, in particular for the protection of this information;
- Change management processes that cover changes to those business processes, systems and services supporting the processing of information;
- Capacity planning to ensure that adequate resources are available to meet operational requirements;
- Back-up process to protect business and customer information from loss;
- Malware protection to protect operational systems against malware attacks;
- Separation of business environments (operations versus development and testing);

- Service management of information security related to service delivery, third party services, supply chain management, contracts and SLAs;
- Operational monitoring and logs, where controls are needed to monitor and record information security–related events;
- Operational software protection, where control measures are needed to protect operational software;
- Vulnerability management, where control measures are needed to be implemented to deal with proper identification, evaluation and resolution of operational system vulnerabilities;
- Operational audits to assess and verify that the operational environment is adequately and suitably protected.

7.4.2 Backups

7.4.2.1 General

There are two main reasons why backups are required. The first, and most important, is to recover information that becomes lost or corrupted due to some system failure, disaster or compromise. Many users and organisations suffer some sort of loss of information at some point in time; it is a common experience shared by many IT users.

Another reason for backing up is for information retention purposes: to enable copies of information to be brought back into the system that were created or processed at some earlier time.

7.4.2.2 Policy

A policy should be produced that defines the business requirements for backing up the organisation's information assets. The slogan "back it up before your business loses it" is very simple way of illustrating this basic requirement. Performing regular information system backups is vital to ensure an organisation's information is adequately protected and preserved from damage or loss. The policy should state the business importance of doing backups and the consequential risks that could severely impact business operations for failing to implement an effective backup process. This should include the risk of exposure to legal proceedings, not being able to respond to customer orders and supplier payments, deliveries being cancelled or production lines being halted or running inefficiently. On the other side of the coin, having a good backup system enables the organisation to operate normally—to continue trading and supplying customers with services and/or products even if the original information has been damaged or lost.

7.4.2.3 Roles and Responsibilities

Information owners are accountable for the safe and secure protection of that information. Of course, the deployment of risk controls to provide such protection may be delegated to others for practical purposes, but the owner still remains accountable. It should be made clear that it is essential that all information that is processed by all departments, operational groups and staff, whether they are owners or custodians of information assets, should be the subject of regular backups.

Much of the information an organisation has and processes is indispensable for it to do its business. Also, the organisation needs to take responsibility for all of the information it handles, even when the information owner is another organisation or its customers/clients. The consequences of loss of information or its destruction or damage can have severe commercial and financial consequences as well as dire legal implications so *back it up before your business loses it* is a clear and simple message.

7.4.2.4 The Technology Is Available; Use It!

Backups can be implemented in many different ways, and today's technology makes the process relatively easy. There many pieces of commercial software and types of storage media (hard disk, magnetic tape, optical disk, solid state, remote storage) that can be used as backup facilities.

7.4.2.5 Managed Backup Services

One option that some organisations consider is the use of managed backup services. There are, of course, advantages and disadvantages with this approach:

- Advantages include that the organisation uses less of its resources implementing and managing backups, it uses less time keeping up with the advances in technology and the maintenance of ICT for backups, it can reduce the organisation's risk of data loss due to lack of focus under operational pressures, human error and/or accidents, it reduces the day to day responsibilities of backups and it provides a way of achieving the automatic availability of offsite storage for added protection;
- The main disadvantage includes that the organisation is entrusting its information assets and the management of the risk to a third party

and not to internal staff or another part of its organisation. the liabilities this entrustment might entail must be weighed, as at the end of the day the organisation is still accountable for these information assets, which could involve information owned by its customers or suppliers and certainly could contain personal data.

If this is the option that the organisation selects, then careful consideration needs to be given to the managed service contract and SLA to ensure the organisation is adequately and suitably protected.

7.4.2.6 Backup Processes

In developing best practice, the following questions should be considered when a process for backing-up information is being decided:

- What are the organisation's objectives for backups?
- Who is responsible for the backup process?
- When are backups carried out?
- How is the storage of the information to be organized and managed? What storage media should be used?
- What are the recovery objectives? For example, what is the desired time objective between the disaster occurring and the recovery and restoring of information in the system?
- What are the performance requirements?
- What are the arrangements for the backup process to be tested to check the validity of the recovery of information?
- Are backups kept secure and offsite from the main business premises?
- Do staff receive some training in doing backups to safeguard their essential information on their own computers, which are not backed up centrally, as well as knowledge on how to use the central backup system as per the backup procedures, if necessary and appropriate?
- Is there a need for backups to be encrypted and/or password protected to ensure their confidentiality, integrity and availability?
- Is a sample of the backup copies archived every two to three months?

Finally always consider and plan for the worst-case scenario, when a disaster occurs and information is corrupted, damaged or lost: how easy is it recover the information?

7.4.3 Capacity Planning

In the context of information security management, capacity planning is the process of identifying the current and predicted future demands of system resources to ensure that the organisation has sufficient capacity to do its business in a productive, efficient and effective way. The objective of capacity planning is to avoid the problems of overloads and reduction of performance levels due to the lack of capacity commensurate with the demands of users and customers. Capacity planning helps to minimize this discrepancy between capacity and demands.

An organisation's information processing and/or networking demands will and do vary due to various reasons:

- Extending the processing operations of the business (e.g., due to increase in sales and production, increase in the customer base, expanding use of supply chains);
- Additions or modifications to the current Internet services it offers and/or its outsourcing services;
- The introduction of new working practices and/or methods of processing such as allowing more staff to work from home and giving them more remote access;
- New computer systems being networked together or new networks being included;
- Redeployment of staff or large increase in the staff count.

The term capacity, for the purposes of capacity planning, can be taken to mean, for example, the amount of information processing power and capability that an organisation needs in a given period of time, the amount of email traffic or the volume of Internet transactions an organisation needs to engage in over a period of time.

Capacity planning is essential to ensure that the organisation has sufficient capacity to meet the demands of business operations and to avoid problems such as lack of resource availability, system overloads, down times or degradation in system processing capability. Capacity planning is an important and critical management control in any risk management strategy and one that needs to be attended to and reviewed on a regular basis. The ISO/IEC 27001 best practice control A.12.1.3 addresses the issue of capacity management.

7.4.4 Change Management

Change management, for the purposes of this chapter, covers changes to those business processes, systems and services that support the processing of information in an operational environment. Our concern here is the impact that any change might have on information security in the operational environment. A change management process should be in place that identifies the change, assesses the potential impact of the change, ensures formal agreement to the change after reviewing the potential impact and takes the necessary actions to implement the agreed change. Add to this that the change management process should take into account any information security concerns, requirements, risks and impacts related to the proposed change and verify that any implementation of the change meets these information security aspects.

Of course, there are many types of change that the organisation might need to consider, either following an internal requirement or as a result of an external requirement, such as changes relating to:

- Business mergers; downsizing or expansions; increases in staff levels;
- New business processes; deploying new ways and methods of working;
- Adoption of new technology; integrating new systems with legacy systems; deploying new network services;
- Third-party service delivery;
- Supply chains;
- Contractual obligations;
- New or revised legislation and regulation.

7.4.5 Third-Party Services

7.4.5.1 Addressing Information Security

More and more organisations are using third-party services for various business functions and applications. The following are a few of these uses:

- Outsourcing call centres and help desks;
- Managed security services;
- Recruitment of staff taking over from the activity of a human resources group;

- Managed data backup and data recovery services;
- Hosting web sites;
- Development of software;
- Recovery of debts; goods relating to unpaid invoices and payments;
- ICT support services.

Whatever the purpose of using external services, it is important that the requirements for information security shall be addressed and be the subject of contractual and service level agreements. At the core of this is the information that is outsourced and processed by third parties and how this information is to be protected. The organisation is giving access to the third party its information assets—information that might be commercially sensitive, operationally critical or personal data. This arrangement is an example of *transfer of risk*. This arrangement does not, however, obviate the organisation's ownership of the information or their responsibility to be accountable for the protection of this information. Having said this, there are many business benefits to using third-party services. The organisation needs to weigh the pros and cons as part of its risk decision-making process.

Considering information security, the organisation should ensure that its information assets are given information security measures that are commensurate with the organisation's business risks and with their business interest.

7.4.5.2 Service Agreements

The main aspects that should be considered with third-party suppliers in agreeing a service agreement include:

- A clear definition of the security requirements and the security roles, responsibilities and obligations of both parties for fulfilling these requirements;
- The process for dealing with information security incidents;
- The access arrangements to ensure the confidentiality and accessibility of the organisation's information assets;
- Backup and business continuity arrangements to ensure the integrity and availability of services and of the organisation's information assets;
- Regular reporting;
- Provisions for management changes;

- Right to review and audit the third party's information security arrangements.

The organisation needs to work with third-party suppliers toward a commercially amicable agreement of how to deal with the information security requirements. Clearly, the legal managers on both sides should be involved. The third-party supplier is more than likely to have standard contracts and service level agreements in place, which may or may not include clauses for the provision of information security management. The organisation and its legal department will need to review these third-party contracts and SLAs to check whether or not they comply with the organisation's requirements.

7.5 Incident Management

7.5.1 Events That Compromise

Information security incidents can cause an organisation's systems to be severely damaged, their work to be significantly interrupted or them to be subject to legal action (both civil and criminal). In the worst case, it might even threaten their survival as a business.

The range of information security incidents is quite diverse and includes any event that could compromise the protection of the organisation's information. This could be an incident caused by someone inside the organisation (an insider threat) or someone outside, external to the organisation. The incident might be caused intentionally or by accident. For example, an inexperienced employee using a badly written procedure, the intention of which is to give instructions on sending file attachments to email messages, might cause the incident. The incident might be a system failure or malfunction due to lack of proper maintenance, or a virus or other malware attack through lack of up-to-date antimalware software. The incident might concern a poor access control system, a badly implemented procedure to prevent ID theft, a flawed process to prevent fraud or a lack of physical protection allowing some unauthorised person to enter a secure area.

The severity of the incident could be minor or major, depending on the level of risk and impact involved. Incidents might occur daily, weekly, monthly and so on due to their uncertainty, so it is important that the organisation has a process in place to handle the incident when it might occur. All incidents should be dealt with in a timely way, but some incidents are more severe than others and they should be given higher priority to resolve. The longer an organisation avoids taking appropriate action, the worse the problem might become and the greater the resulting cost—timeliness is critical to protect the organisation.

7.5.2 Use Cases

7.5.2.1 External Attack Hits a Auto-Industry Supply Chain

A medium-sized auto-engineering supply company is experiencing a number of external unauthorised attempts to its internal networks. None of these were successful, but management was worried by this concentrated spate of attacks. Their firewall detecting then followed this and blocked a rapid increase in the number of malware-based attempts to penetrate their system. At the same time, the company's email system was being bombarded with a continuous stream of thousands of unwanted emails—enough to overload its system. This chain of events further worried management, as it seemed that the company was being targeted. Fortunately for the company, its incident handling process was able to deter any serious impacts.

It turned out after consulting with one of their end customers that one customer in particular was the main target, and all the customer's suppliers were indirectly targeted to interrupt the supply chain. The incident involved the customer's supply management system being compromised. After many weeks of investigation involving the end customer and three of its suppliers, several men were charged with criminal offences, including employees from all the companies involved. The impact on all companies involved was costly and resource intensive. The media coverage reporting also had an impact on the company and its suppliers.

Finally, over a period of time, the situation got back to near normal and the company started building back the confidence and loyalty of its customers. All those involved learnt many lessons in handling this supply chain incident and should help them in avoiding future incidents such as this.

7.5.2.2 Down Tools on the Production Line

A medium-sized company producing a range of electronic components is hit by a spate of system problems. Their design and production line are highly dependent on the use of CAD/CAM applications for design work and automated assembly and testing facilities. The processes seemed to become erratic, unreliable and produce faulty results, which slowed down productivity and on two occasions the complete system had to be restarted.

The company's incident team was assembled to resolve the problem. In collecting all relevant information about these incidents, they discovered the problem seemed to be related to few new customized application software packages that had been deployed two months ago but had never been tested before being used in an operational environment. The supplier of this software was immediately contacted and asked to solve this problem. Over the next 24 hours, software engineers found a number of bugs in the software products and produces several patches to resolve the issue. The customer tested these and found them now to be functioning as intended.

The company got back into to its normal operational working and productivity after three days of down time. Since then, several improvements have been made to its information security and software management controls.

7.5.2.3 Online Services Company

The Online Services company has recently introduced a strict Internet Usage Policy to improve staff productivity and customer service, and to direct staff to spend less time surfing the Internet for personal reasons during busy times in the office. This seemed to work for a short period of time, but then a small group of staff became disgruntled and at odds with management. Shortly afterwards, the company seemed to experience an increase in the number of unauthorised access attempts during the first two weekends in March. There was also an increase in the number of internally sourced blocked viruses.

On inspecting the Internet usage records, the company noticed that a pattern between blocking external access attempts and internal access attempts. More detailed analysis of other monitoring records kept by the organisation showed that those attempting internal access were those same people that expressed their strong objection to the recently introduced policy, and three of these worked during the weekend of the 3rd and 4th March.

Company officials interviewed the staff and then started disciplinary procedures. During the process the names of friends who were involved in the external attempts finally emerged. It was also discovered that some of the unexplained loss and damage of company files could be traced back to the activities that took place over the weekend of the 3rd and 4th March.

The relevant authorities were called in to help in the case and to make charges where appropriate. This group of collaborators, staff and friends were responsible for gaining unauthorised access, destroying files and planting viruses in the company's system. The enquiry concluded with internal staff and external individuals being charged by the police for criminal activities.

7.5.3 Processes

The general process of information security incident management is illustrated in Figure 7.1.

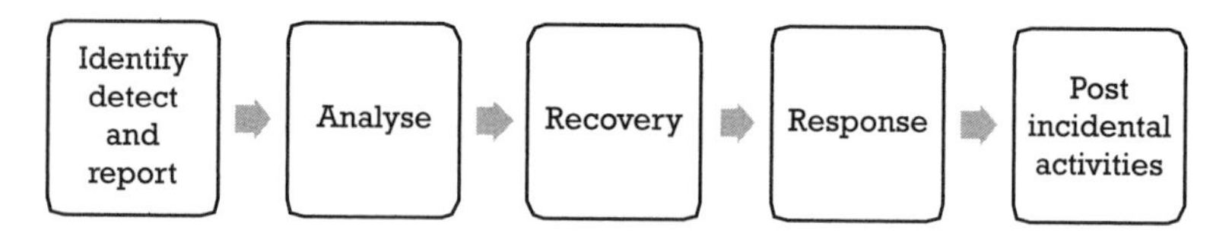

Figure 7.1 Incident management process.

7.5.3.1 Planning

Being prepared is vital to being effective in the event of a major incident. This is a fairly common sense thing to say, but taking the right action toward being prepared is not often done in practice. In many cases, producing a set of procedures for incident handling and having the procedures tested and tuned to the organisation's needs more often than not takes place only after a few major incidents have happened. Some organisations partially address the problem (e.g., they may have an incident system in place but nothing else to protect against malware or software vulnerabilities and not a comprehensive system that can deal with any form of incident).

So it is important to have in place a set of policies and procedures defining a typical set of incidents; the process for identifying and evaluating these; an agreed response, resolution and recovery process; a "what next" process; a determination of who is responsible for what part of the process; and the means of communications during an incident while it is occurring. It is important that this is not just a paper exercise, but that these processes are tested in advance of any potential incident occurring.

The planning activity should establish an incident response team with defined roles and responsibilities. This planning activity should also define whom to engage with regarding incident events: internal groups (senior management, security team, IT support, legal team, media relations, human resources and so on) and external groups (affected external parties, software vendors, CERTs, ISPs and so on).

7.5.3.2 Identification, Detection and Report

The process needs to identify, detect, report and record:

- What the incident is (is it minor or major)?
- Where in the system is it occurring (internally, externally or both)?
- What resources are being affected (internal and/or external)?

There may be a precursor, a sign, that an incident may occur in the future, or an indication that an incident may have occurred or may be occurring now. Accurate reporting and recording of this information on an incident reporting template is essential to the analysis and responding phase of the process.

7.5.3.3 Analysis and Evaluation

The information that is reported needs to analysed and evaluated by the team to assess the extent and scale of the incident, the severity of the incident, and to ask questions such as:

- Is what is being reported an actual information security incident or some other type of event?
- Is the information accurate? Does the information reported provide a complete picture of the situation? Some indicators may provide information that is incorrect or inaccurate, incomplete, misleading, ambiguous and introduce false positives or false negatives.
- What is the nature of the incident? What are its characteristics? Are the symptoms clear, obvious, and easy to understand?
- What is the scope of the incident? The scope should indicate which parts of the business are affected; what information systems and business processes are affected; what networks, IT systems, applications, services are affected; and the who, what and how (i.e., source and cause of the incident and how it is occurring).
- Is it contained in one area or is it spreading?
- If it is spreading, how fast is it spreading?
- Is it isolated to just internal systems, has it come in from the outside or is it spreading to outside?

These and many other questions need to be asked to be able to start the response and recovery process knowing what should be down, where and to what. Also both internal and external stakeholders need to be notified.

The incident response team needs to analyse, check and validate each incident. On deciding whether the event is actually an incident, the team should do an initial analysis to determine the scope of the incident and sufficient information to prioritise what actions need to be taken and what response strategy to take (e.g., containment, eradication).

7.5.3.4 Response

In responding to the incident, the team needs to work in a timely way and to recover the system to its original operational state as soon as possible. Time is of the essence to stop the problem spreading and to ensure the organisation can keep functioning properly in order not to damage or lose its business through excessive delays caused by the response process.

Some key points:

- Prioritisation of what to deal with first to respond to the incident is of crucial importance; expedient management decision making is critical to the success of the response, resolution and recovery. The incidents to be dealt with as high priority will depend on the business impact and recoverability requirements; operational impact; information security impact; corporate, financial and legal impact; and recoverability—but, of course, people's safety and security should get the highest priority, above all else.
- The nature, severity and spreading of the incident and the resources affected determines the time and effort needed to get the business operations, processes, systems and networks back to normal. The time and resources to recover may be extensive.
- Get the information security handling team started on dealing with recovery.
- Process the evaluated results and evidence about incident.
- Take action to contain the incident.

7.5.3.5 Timeliness of Response

An incident can be thought of as the coming together of a threat and vulnerability. When a hacker is able to exploit the weaknesses in an organisation's access system software or an access method, the result is an incident of unauthorized access. The impact can vary depending on what the hacker is able to do and how fast the organisation is at stopping the hacker. In the case of malware, this depends on how fast the organisation is able to stop the spread of the malware infecting IT systems. Therefore, time is of the absolute essence in dealing with incidents, especially that are severe and involve critical business systems or processes.

The hacker may start by getting access to files and downloading them, then installing malicious code in the system, and finally using the organisation's network vulnerabilities to connect to the network of one of its major customers. This whole chain of events, if allowed to happen, could result in various financial and legal impacts as well as loss of its image and reputation. If this chain of events is contained early enough in its cycle, then the damage and impact to the organisation can be limited. So timeliness of response is very important.

Of course, if the incident puts people's lives at risk, their safety and security is the highest priority, but again a timely response is of the essence. Such might be the case with an environmental incident, an explosion, a bombing or a natural disaster. The people safety aspect takes highest priority over everything else.

7.5.3.6 Recovery

Recovering from the incident involves getting the system that was affected back to its normal operational state so that business can continue to operate. Again, time is of the essence to avoid any unnecessary losses to the organisation during the time when the incident was active and the organisation was not fully operational. Some key points:

- Contain, eradicate and resolve the problem.
- Restore.
- Reestablish any affected computer systems and other IT equipment, computer networking capability and connections. Recovery may involve:
  - Restoring from backups any lost, damaged or corrupted data that occurred during the incident;
  - Rebuilding systems;
  - Replacing files that have been compromised;
  - Installing patches;
  - Changing passwords and accounts and improving system and network security;
  - Reconfiguring systems and networking;
- Reestablish and restore business systems, processes and services back to a normal operational state—check that everything is functioning correctly and as expected; communicate this to staff and management, customers, suppliers and other appropriate stakeholders.
- Fully document the incident case and changes that need to be made to the system, policies, procedures and processes; undertake any final actions to close the incident case file.

7.5.3.7 Post Incident Review and Follow Up

This part of the process is important—first, to do a review of the incident situation, and, second, to take preventative actions to make the necessary improvements to protect the organisation from further incidents of this type from reoccurring.

Post incident review:

- Review the incident report forms;
- Carry out a forensic analysis of the incident;
- Review the effectiveness of the incident handling process and the information security incident handling controls;
- Undertake a business impact analysis of the incident.

Improvements and updates:

- Make recommendations for improvements to the ISMS;
- Take steps to improve the incident management process and controls;
- Update policies, procedures, processes and training where appropriate;
- Where necessary make changes to contracts and SLAs;
- Finally produce an incident report.

7.5.3.8 Incident Investigations

Incident investigations involve the use of digital investigation techniques to identify, gather, evaluate and preserve information (digital evidence—information stored or transmitted in digitak form that may be relied on as evidence) related to an incident that has occurred. Such information might be used as evidence, for example, in the case of civil or criminal actions or for internal management purposes (e.g., internal policy breaches and violations, resolving operational problems and system recovery).

Examining and investigating (as mentioned in ISO/IEC 27001 Annex A.16.1.7) a "collection of evidence" is covered in more detail in the standards referred to in Section 8.6.7.

7.5.3.9 Incident—Cause and Effects

The situation an organisation faces is taking action today based on what happened in the past and its current experiences of the present—yesterday's causes become the results of today (the effects), and tomorrow's results are caused by today's events. The causes and effects are both events—the cause of the security compromise yesterday has an effect on the organisation today; the effect of not resolving this problem today causes continuing or a set of new problems tomorrow and so on. We should learn from the mistakes of the past that cause our security problems to help us do the right things today and to protect our future—see Figure 7.2.

7.5.4 Incident Management Team

An information security incident management process involves those activities to identify, analyse and correct problems and facilitate the response and recovery of the systems that have suffered from the incident. This process should involve a multidisciplinary team, with a range of skills needed to deal with a range of issues. Typically, the incident might require the involvement of management, information security staff, IT and non-IT staff, physical security staff, human resources, legal advisors, those handling external relations and the press, and sometime external parties emergency services, policy and maybe even trade unions.

The team needs to be able to analyse the situation, determine the scale and severity of the compromise, initiate appropriate action to correct the

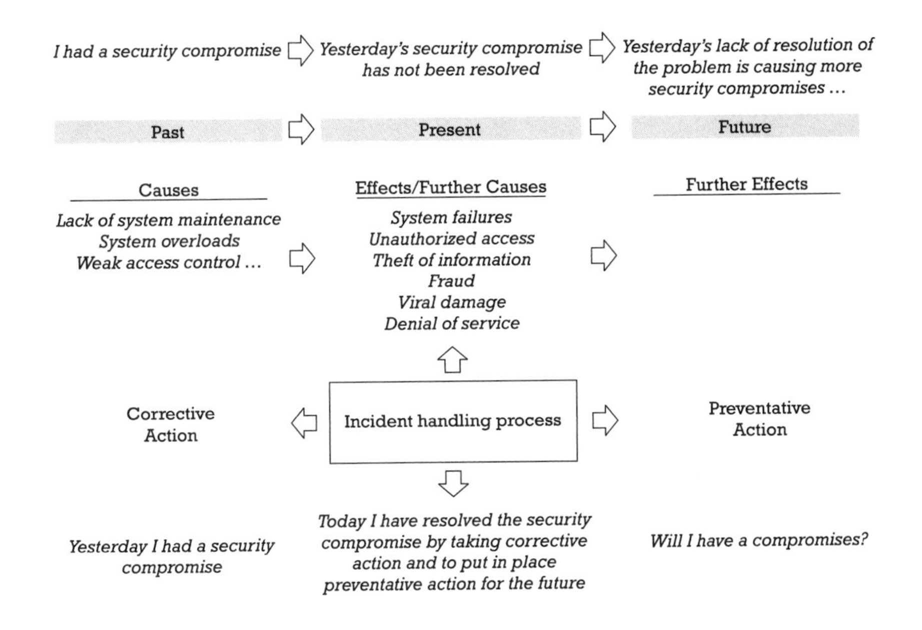

Figure 7.2 Incident—the cause and effect.

problem and prevent it from reoccurring and liaise and communicate all the relevant internal and external parties. The incident handling team should have a leader whose task it is to coordinate the process and facilitate and manage the team that is deployed in the process.

7.5.5 Standards

More detailed technical information on information security incident management can be found in the following sections.

7.5.5.1 ISO/IEC

- ISO/IEC 27035 (Information security incident management):
  - Part 1: Principles of incident management;
  - Part 2: Guidelines to plan and prepare for incident response;
  - Part 3: Guidelines for incident response operations.
- ISO/IEC 27037 Guidelines for the identification, collection, acquisition and preservation of digital evidence;
- ISO/IEC 27038 Specification for digital redaction;
- ISO/IEC 27041 (Guidance on assuring suitability and adequacy of incident investigative methods);
- ISO/IEC 27042 Guidelines for the analysis and interpretation of digital evidence;
- ISO/IEC 27043 (Incident investigation principles and processes);
- ISO/IEC 27044 Guidelines for security information and event management (SIEM) (still in development);
- ISO/IEC 27050 Electronic discovery (still in development);
- ISO/IEC 30121 Governance of digital forensic risk framework.

7.5.5.2 NIST

- SP 800-61 Computer Security Incident Handling Guide;
- SP 800-83 Guide to Malware Incident Prevention and Handling;
- SP 800-86 (DRAFT) Guide to Applying Forensic Techniques to Incident Response.

Note: This list of standards and guidelines is not exhaustive and there are many others available from different sources, including standards bodies, industry associations, user forums and computer emergency response organisations.

7.6 ISMS Availability and Business Continuity

7.6.1 Value and Importance

There should be little doubt as to the importance and value of ISMS availability, information security incident management and business continuity play in supporting the organisation.

We have many global examples of business systems, processes and services failing and being inoperable for hours, crippling businesses whose operations come to a halt. The causes of these incidents can range from natural disasters to online attacks with the deliberate intent of bringing systems down. Today's business reliance on the Internet and IT should place more of an emphasis on businesses being ready to respond to such incidents and to recover as soon as possible to ensure an acceptable level of continuity of business operations. The business impact of not having appropriate resilience, response, contingency and protection processes in place can all too well be seen by the various real cases that have hit the headlines, from man-made situations to natural hazards or a combination of both (i.e., where a natural hazard triggers a disaster relating to some man-made situation). Headline news raises awareness of how vulnerable our systems can be and the impact on business when things go wrong. This heightened awareness draws the attention and concerns of organisations, citizens and consumers alike. It triggers some organisations to rethink how would they cope if it happened to them. They need to question and examine how good their processes and systems would be at coping with these issues.

Many companies have not sufficiently protected themselves with an effective business continuity process, and some do not even have such a process in place. Sometimes the reason that business continuity is not in place is that organisations believe it presents too many challenges and/or is too expensive or they just don't know what to do. These are some of the reasons why small-sized companies do not go for business continuity, as the cost, know how and resources are out of their reach. This, of course, should not be the case, since there are many routes available for getting simple advice on what can be done, and the implementation of some of the basics should not be out of the reach of small-sized companies. At the end of the day, the company needs to weigh the costs of the resulting impact of not doing anything against the cost of doing something with the reduction in the business impact. It could make the difference between survival of the business and closure of the business. It should be clear that there is business value in doing it, no matter how perfect the solution is. It is important to take that first step of doing something, to build upon the basics, and the rest will fall in place with the right guidance. One of the important things to do first is to get management commitment, and this can be achieved by doing an initial business impact analysis.

7.6.2 Business Impact

One of the important questions to ask is, "What are the critical things that are essential to the organisation to enable it to continue to operate?" Once we know this, we can then test these out by looking at some examples that might appropriate to the organisation running its business. These can then be used to calculate the business impacts the organisation might be challenged with in the event of disastrous incidents occurring.

If the organisation is not prepared and has no continuity plans in place for the advent of a disaster, disruption to its operations, a major systems failures or external influences from the markets it's involved in, then it leaves the organisation wide open to being exposed to high impacts that could ruin the business. For example:

- Supply chain failures and disruptions;
- Denial of service attack shutting down or grossly restricting operations dependent on ICT systems, services and a networking capability;
- Critical loss of supporting infrastructure (e.g., electricity, gas or water supplies) and services (e.g., telecommunication services, ISP services, outsourcing facilities);
- Major economic failures and losses, including critical depression in market conditions (e.g., share prices, interest rates, world/country economies);
- Strikes, political activities or breakdowns in workforce relations;
- Environmental hazards and disasters;
- Bombs, explosions and terrorist attacks;
- Major legal, regulatory or contractual failures leading to penalties and liabilities.

These examples cover financial, operational, customer/supplier, staff-related and legal issues.

7.6.3 Plans

It is essential that the organisation should have in place a business continuity plan. This plan should include documented procedures that will give clear instructions of what to do to respond, recover, resume and restore to an acceptable level of operations following a disruption, disaster, emergency or failure. This plan should include information on the required resources, services and activities to be undertaken to ensure the continued use of critical business functions, processes and systems.

The scope of the business continuity plan is likely to cover a broad range of topics and activities, especially providing availability, continuity and recovery of critical business elements: business functions, processes and services. The scope of the plan should cover aspects of disaster recovery, incident management, information security management, safety management and emergency and crisis management. ISO 22301 provides more details regarding continuity plans and ISO/IEC 27002 Chapter 17 covers the information security aspects of business continuity management.

7.6.4 Processes

Business continuity planning and supporting processes are aimed at ensuring that the right resources are available and the right activities are being undertaken to keep continuity of the critical elements of the business running and operating, and if these critical elements become unavailable and inoperable during a disaster that processes are in place to recover these as a matter of urgency.

In today's business environment, where ICT is a vital tool to the proper functioning of the business, it is important to protect the ICT accordingly in the event of a disaster. When the ICT has failed, been compromised, become damaged or become unavailable during a disaster, then timely recovery of the ICT is vital, which might mean replacing damaged ICT.

An essential process included in business continuity management is undertaking an impact analysis. This involves gathering, analysing and evaluating this information to assess the impact with regard to loss of confidentiality, integrity and availability. The results of this can then go toward a review and decision-making process to determine the organisation's requirements and what should be done to guard against such business impacts occurring.

For more detailed information the reader is directed to ISO 22301 and ISO 22320 (see Section 8.6.5).

7.6.5 Standards

More detailed technical information on business continuity and information security incident management can be found in the following sections.

7.6.5.1 ISO

- ISO 22301 (business continuity management system—requirements);
- ISO 22313 (business continuity management systems—guidance);
- ISO 22320 (emergency management—requirements for incident response).

7.6.5.2 ISO/IEC

- ISO/IEC 27002 (Chapter 17, Information Security Aspects of Business Continuity Management);
- ISO/IEC 27031 (guidelines for ICT readiness for business continuity);
- ISO/IEC 27035 (information security incident management).

7.6.5.3 NIST

- SP 800-34Rev1 Contingency Planning Guide for Federal Information Systems.

7.7 ISMS Use Examples

7.7.1 SME Design Services

7.7.1.1 Introduction

A small company with 11 full time staff members provides household designs for domestic users. This includes customised fitted kitchens and bathrooms, studies, living rooms and dining rooms. The company consists of consultants and designers, salespeople, showroom staff, the owner and a secretary.

The company has been in operation for seven years and is well recognised as a supplier of quality designs. The company has just one showroom/

office located on the ground floor of a five-floor building on a busy high street of a town in the UK. The company has already been operating an ISMS for the past three years. They now want to expand the business to network its PCs, give remote access via consultants' laptops when on customers sites, add web site services to its customers and open two other offices in other European countries to provide the same sort of services.

The company has carried out a SWOT analysis to assess the feasibility of the objectives and potential success of expansion. Based on this analysis, it decides to go ahead. Their business implementation plan includes an extension of the scope of its ISMS to include its website presence, remote access and its two new planned offices.

7.7.1.2 Existing and Future ISMS Deployment

The company went through a gap analysis to identify that their existing information security system includes:

- An information security policy;
- Procedures on backups, viruses, and handling personal and customer information;
- Password control on all office PCs and consultant laptops;
- Physical access controls to its offices, rooms and filing cabinets.

A target ISMS scope is defined to cover all existing operational needs as well as future plans for expansion. A risk assessment on the new ISMS scope is done, taking into account the increase in its resources and the expansion of its business processes, the results of which are evaluated and developed into a risk treatment and ISMS implementation plan. As their future business plan rolls out over the coming six months, so does the implementation of their ISMS, and within a further six months they are fully operational. Their Internet side of the business starts to grow, as does the number of staff employed by the company, as they are getting more customer enquiries and orders.

7.7.2 Legal Services

7.7.2.1 Introduction

A small legal firm with 17 full-time staff members provides legal advice for both commercial businesses as well as citizens. The firm has three offices in different cities in the UK. It has been established for more than 25 years and has many high-profile corporate clients on its books. Over the past three years, it has relied more and more on the use of IT services and the Internet to conduct its business. It is now becoming quite clear that it needs to do something about enhancing its existing information security to take into account its high-tech way of doing business.

7.7.2.2 Existing and Future ISMS Deployment

The firm already has security measures in place for its business, but these are stand-alone solutions with very little networking deployed. The security it has in place needs to be enhanced to meet the needs of management for a more integrated, networked environment. The decision to use ISO/IEC 27001 was made after the firm was made aware of other firms in the legal services business going down the same route. Its main IT resources and Internet capability consists of:

- A number of office PCs networked together to a central server;
- Internet access via a firewall connected to the office server;
- Laptops, tablets and mobiles (all Wi-Fi enabled) used by legal professionals.

The information it processes includes:

- Confidential client case files, reports, letters, statements and other documents;
- The firm's internal confidential and restricted information, documents and files;
- Library of legal books, works of reference and other generally publicly available information.

The task they embarked on (with the help of external consultants) was first to identify the needs and expectations of management, clients and other interested parties relevant to an ISMS, then to define an appropriate ISMS scope commensurate with the nature of their business, before carrying out a risk assessment and risk treatment exercise. The risk assessment and risk treatment exercise concluded with the need to implement a range of new controls to enhance their existing protection and to facilitate protection of their newly networked business environment. This led the firm to implement an ISMS to the requirements of ISO/IEC 27001 and for an audit to be carried out by an accredited certification body, which resulted in the award of a ISO/IEC 27001 certificate.

7.7.3 Electronic Accounting System

7.7.3.1 Introduction

The management of a large retail company decides to install a fully integrated electronic accounting system, but before doing so the decision is made to check the requirements for achieving an effective level of information security. The current accounting system does use PCs, but these are not networked together to provide a streamlined start-to- finish process for handling sales, purchases, orders, invoices, payments, delivery and dispatch notes. Their future plans require the installation of an IT system networked throughout the company with links to external customers and suppliers, with appropriate software application packages. This will enable staff to process and share information online within the company and also to electronically process orders, payments and other documents to suppliers and customers.

7.7.3.2 Existing and Future ISMS Deployment

Although the company has protection measures in place, it decides to carry out a complete overhaul of its information security taking account of existing controls. They recruit someone as their security officer and together with departments such as sales, purchasing, finance, operations and IT services they develop an ISMS scope and design based on ISO/IEC 27001.

They get management approval and commitment to implement this ISMS design across all of its five locations. Although the project is quite a sizable undertaking the management team wants to achieve this electronic accounting system development as soon as possible to take advantage of the business benefits of such a system and provide added value to their customers, they decide to supplement their internal project team with external consultants.

In addition the gap analysis and risk assessment was not easy to accomplish due to multisite nature of the project; the current lack of coordination, consistency and integration of the current system; and some local resistance from staff regarding issues to do with change and questions on security. It is then agreed that senior management must get more closely involved to show their full support and to commit to more resources. After several months delay in making progress, the company rolls out a fully operational electronic accounting system with information security in place. Due this experience, management is persuaded to recruit a second full-time employee to support the security officer to maintain the ISMS. Over the next six to nine months, the company prepares itself for third-party certification, which it achieved.

7.7.4 Government Payment System

7.7.4.1 Introduction

An e-government department in Southeast Asia operates both an offline and online payment system refunding citizens for their social security claims. As with many governments around the world, the attraction of converting their business processes to utilize e-commerce technology and providing the public with online access to public information and government services (e.g., to pay taxes, claim benefits, apply for visa applications and so on) is strong. Such use of e-commerce can be used for internal government business, by government for external organisation business and transactions, or by government for citizen transactions and sharing of information. In this scenario, another government department whose responsibility it is to advise on information security matters recommends that all e-commerce applications go down the ISO/IEC 27001 ISMS route.

7.7.4.2 Existing and Future ISMS Deployment

This particular government department, dealing with social service payments, always had information security measures due to the sensitive nature of the claims they were handling, especially as they involved general personal data as well as more specific details such as health-related information. Their future plans had covered moving over to a more electronic way of submitting claims with "social services Internet café"–type facilities in government centres and offices around the country all networked to the department dealing with social services. The ISMS for this social services Internet café had to deal with citizen privacy technology, citizen smart cards for identification and authentication purposes as well as accessing the system, secure exchange of citizen information and secure information processing facilities in the government department. Of course, around this is the management framework for making this function effectively.

The scale of this project is large and complex, and so it is decided to roll this out in phases, focusing on one particular town of a medium-sized population to trial out the ISMS design, to improve the design where necessary and then to proceed with other towns to gradually cover the whole country. This first trial was successful, although ISMS did need various adjustments and improvements regarding its implementation controls. After subsequent phases of work had been completed, growing the ISMS solution to more than 70% coverage of the population, the government decided to go for a third-party certification audit.

7.7.5 Outsourcing Call Centre Operations

7.7.5.1 Introduction

In the interests of streamlining the business and cost cutting, a major international European-based company outsources its call centre operations to a company in Asia. This type of business has become very common and popular over the years, and one that is expected to grow. Their outsourced call centre operations cover the company's customers all over Europe, and the daily number of customer enquiries continues to grow due to their business successes.

7.7.5.2 Existing and Future ISMS Deployment

It is of utmost importance that an information security risk assessment is carried out before undertaking such an operational change, especially to consider the various factors involved in outsourcing to overseas countries, including the following.

Geographical, Environmental and Political Factors

- What hazards apply to this particular area of the world? Is it in an earthquake zone or an area prone to volcanic activity, tornadoes, typhoons, tsunamis and other natural phenomena? Many areas in Asia have these types of problems.
- Is the location in an area of political and social stability? Is it economical stable?

IT Facilities, Communications and Levels of Service

- Are the communication systems able to cope with the volume of customer traffic?
- Can the outsourced company supply adequate facilities and resources to fulfill the contract requirements (e.g., sufficient capacity to meet demands and effective information security, backup, disaster recovery and business continuity arrangements, response times, performance and service levels)?
- Can the location and IT facilities be easily secured from a physical perspective from local threats?

Information Security

- Has the outsourced company carried out a risk assessment and risk treatment exercise? Has the outsourced company implemented adequate information security to protect the information of its customers? Does it provide appropriate information separation of each of its customers? Does it take regular audits and reviews of its information security? Does it have an incident management process in place? Does it have a business continuity management system?

Staffing Levels

- Can the outsourced company supply an adequate number of trained staff to fulfill the contract requirements (e.g., communication skills, knowledge, experience and language skills)?

In the case of this particular outsourcing organisation and its information risks, it has developed an ISMS, it is being used to provide protection of its call centre operation and it has been certified to ISO/IEC 27001 to demonstration to its customers that they are fit for purpose. Sometimes, it is the customer itself that requires the outsourced company to be certified to ISO/IEC 27001. It should be noted that sometimes, it is the customer itself that requires the outsourced company to be certified to be in conformance with the requirements of ISO/IEC 27001.

7.7.6 Manufacturing Systems

7.7.6.1 Introduction

A successful medium-sized manufacturing business is certified to ISO 9001, or quality management system (QMS). It has now decided that it should consider applying ISO/IEC 27001 across its business to complement its ISO 9001 capability. It would eventually like to integrate the two systems together to create an efficient and effective management framework to gain the cost benefits of having an integrated certification audit.

7.7.6.2 Existing and Future ISMS Deployment

Its current information security practice seemed to have been sufficient in past years, but there is growing awareness of the threats and risks the company faces from observing market trends in its particular industry sector. An additional drive toward gaining more awareness is the fact that one of its competitors suffered a damaging attack on one of the competitor's IT-based automated production lines. The company employed a consultant to do a gap analysis and risk assessment of its current systems. The results of this were quite startling for the company, as it turned out its current information security provisions were only 35% conformant with ISO/IEC 27001, and in fact 20% of these had been picked up by the recent ISO 9001 surveillance audit as things that had an impact on its QMS implementation. Even more startling was the large areas of noncompliance, 55% of which turned out to be high-risk areas based on the results of the consultant's risk report. Their main areas of risk were:

- General lack of management policy and procedures;
- No operational procedures in a number of key areas, such as for information handling, backups and access control;
- No regular risk assessment programme;
- No regular monitoring, audit or reviews of its information security;
- Insufficient access control for operating systems, internal networked services and applications;
- No information security incident management processes;
- The gateway to the Internet was wide open to external threats, allowing relatively easy access to the internal network;
- Lack of information security awareness and training;
- Insufficient resources to carry out information security–related tasks;
- Lack of information security considerations in its contracts and SLAs with third-party service providers;
- No vulnerability management;
- Lack of protection against social engineering attacks;
- Lack of information security for business continuity.

After several management decision-making meetings, the company decided to go ahead with developing an ISMS for the whole company. The work was contracted to a small company specializing in this work, which had a track record in different parts of the world in helping clients to establish ISMS solutions in conformance with the requirements of ISO/IEC 27001. After an intensive contracting period, the company implemented an ISMS. The company even went a stage further to get certified to ISO/

IEC 27001, out timing their competitor. Its next challenge was to work with the certification body toward planning an integrated ISO 9001 and ISO/IEC 27001 audit.

7.7.7 Supply Chain Management

7.7.7.1 Introduction

Supply chains are the norm for most organisations. For the organisation to work efficiently, it needs to manage its operational and business relationship with its supply chain effectively, especially if the business is in a fast-moving and dynamic market where market competition, customer satisfaction, timeliness and profitability are vital. The supply chain can be viewed as a networked system of organisations, human resources, operations and communications and information and IT resources all working together to deliver a product or service to the end customer. The supply chain industry thrives on the flow of accurate information being available 24x7 (i.e., 24 hours a day, seven days a week).

Typical examples are online purchases of books, food supply chains, car parts, components and subassemblies to support car manufacturers, energy supplies, supply chains supporting the construction industry, supply chains supporting the travel industry or supply chains supporting the production and supply of medicines for the health care industry. All of these supply chains need to run both efficiently and effectively to fulfill customer and consumer needs and demands.

Supply chains link together an organisation's value chain of activities, such as the logistics of both its inbound and outbound services, its operations in producing things or delivering services, its marketing and sales support and its maintenance support.

Outsourcing companies (sometimes referred to as third-party logistics providers, or 3PL) are available to support an organisation's supply chain, such as in the provisioning of integrated warehousing and transportation services.

7.7.7.2 Existing and Future ISMS Deployment

Again, supply chain management is another application where ISO/IEC 27001 is able to provide a management framework to manage its information security risks. Some organisations that supply logistic support services have been certified against ISO/IEC 27001 for this very purpose, especially those that also provide international logistic support.

CHAPTER

8

Contents

Performance Evaluation

8.1 Performance, Change and Improvement

8.1.1 How Effective, Adequate and Suitable Is the ISMS?

Is my ISMS fit for purpose? What exactly does this mean? Here, we are taking it to mean that the performance of the ISMS is effective, adequate and suitable to meet with the organisation's policy, strategy, objectives and business and legal requirements. This leads to the question of how do I evaluate the ISMS to check its performance, and the answer is by monitoring aspects of the ISMS, taking ISMS measurements and reviews. So are the ISMS processes successfully providing protection of the organisation's information resources? Are the policies and procedures suitable? Are the ISMS resources adequate to manage the risks? These and many other questions arise when we start to evaluate the performance of the ISMS to check that it delivers the desired and intended result (i.e., the ISMS addresses all the requirements and issues discussed in Chapter 3; it is managing the risks as discussed in Chapter 4; it has implemented ISMS with suitable and adequate policies; processes, procedures and other controls to manage the risk as discussed Chapters 6 and 7; it is monitoring and reviewing the ISMS) as well as the requirements in Chapters 5 and 7 on leadership and mangement support and it engaged in continued improvement (Chapter 10).

Performance levels of the ISMS will vary as time goes by because of change: business changes, market changes, political and economic changes, resources changes, environmental changes, legal changes, technological changes. These are a few of the types of changes that the organisation may face, and all of these carry with it the chance that the levels of risk and impact will change.

8.1.2 Change and the Certainty of Change

"Nothing is as constant as change" is an old adage, which is very apt. It is often said that change is the quality of impermanence, and everything on this earth is impermanent, in a state of constant flux. The Greek philosopher Heraclitus saw change as ever present, ever flowing and all encompassing. Change is at the core of continuous improvement and making sure the organisation is managing its risk to ensure it's protecting its information at all times. Monitoring and reviewing changes, the causes and effects enable the organisation manage the resulting changes in risks and impacts appropriately.

Changes may happen slowly or rapidly, decrease or increase, go forward or reverse or progress in the same or different direction; everything at one point in time will change, whether it is something that is physically strong or weak, tangible or intangible. Change can be sequential, nonsequential or consequential. Organisations need to view change as a natural process, and they should monitor and review the patterns of change, the speed and direction—its velocity—and any other aspects that will affect and impact the organisation.

As Sun Tzu's words reflect that the organisation knowing itself and its business, and knowing the environment and markets they are doing business in, can lead to a better chance of achieving 100% success. Of course, continuation of this success requires the organisation knowing, understanding and adapting in a suitable way to both the internal and external changes. The organisation knowing its information security strengths and weaknesses (vulnerabilities) and also the threats that might exploit these gives it a better chance of managing its information security risks. Regular monitoring and reviewing any changes to these strengths, weaknesses and threats enables it to be in a better position to continue to improve and maintain an effective information security capability.

An organisation knowing its online business capability and its information security strengths and weaknesses, and its competitors' capabilities, strengths and weaknesses in this field, enables the organisation to manage and improve its information security risk controls capabilities. Again, regular monitoring and reviewing of any changes to these conditions to continue is necessary to improve and maintain an effective information security capability and enable business opportunities.

How much change has occurred? What are the effects of the change? What is the level of change in risk and impact? These and other related questions all become the subject of measurements as part of the overall performance evaluation exercise.

8.1.3 Change Management

8.1.3.1 Change Happens

There are two basic forms of change: change that is planned by the organisation (an organisation plans to adopt a new technology) and change that is unintended or unplanned by the organisation and not under its control (changes in the market, legislation that the organisation needs to respond and adapt to). In both cases, information security is a critical element in adapting and/or responding to the change.

Change management is about the path of transition from the current state of the business operations to the future state of business operations. The process of transition can go relatively smoothly, or it might be fraught with difficulties. As such, change management involves members of staff, business and operational teams, management and the organisation itself. How an organisation adapts to the change process can make a big difference to its business. If there is resistance to the changes, then this can hamper the overall success of the transition and the end result. Individuals in general may be anxious about change and might have an innate sense to resist. In the case of major business change (e.g., downsizing, a merger, more automated business processes), the organisation needs to help its staff to prepare and make the transition as smooth as possible whilst helping to allay any anxieties.

There will always be individuals or certain business groups, rather than the whole organisation itself, who resist change: this situation is normal. Taking the path of least resistance is a good strategy for many difficult situations in all walks of life. Using an area of the business it knows to be open and receptive, and that offers the least resistance to a proposed change, is a good starting point to introduce and test out the change. Once this area of the business gains confidence through deployment of the change and other parts of organisation see its success and value to the business, the organisation can then start to roll out the change across the rest of the organisation.

It has been said that the key to success is to keep change to a minimum and not go for change for the sake of change. Some things may not, or should not, need to be changed, in order to ensure consistency and business continuity. Adoption of changes should be in flow with the business and going for a "big bang" approach is not necessarily the best move; it is better to change strategy via a path of least resistance. Of course, it is entirely up to the organisation, it is its commercial choice how much change is necessary to meet its future mission and business objectives. Whatever set of changes are proposed by the organisation, an information security risk and impact assessment needs to be carried out before any changes are implemented. For example, a complete overhaul and upgrading of the organisation's IT strategy for both IT acquisition and deployment needs a good knowledge of the internal and external factors that might have an impact on this strategy. In addition, the information security controls required in relation to the use of this new/upgraded IT needs careful consideration (e.g., knowing what external changes have driven this new strategy, what the changes will mean to its staff as well as to its customers and suppliers or what impact these changes will have on its market positioning).

8.1.3.2 Process

Change management processes must consider:

- How the changes fit into the bigger business environment the organisation operates within, making sure the implementation of the changes work for the organisation;
- There are no conflicts with the ISMS implementation, and all the necessary risks and impacts have been taken into account;
- There is adequate awareness and training of staff with regard to the changes, including any training related to information security;
- Communication of changes to all interested parties;
- Gaining the support and cooperation of staff for the adoption of the changes.

8.1.3.3 Types of Changes

There are several types of changes, including those driven by external forces to which the organisation needs to decide how to respond and react, and those planned and instigated by the organisation (e.g., where it can be more proactive in driving and leading the market with new customers, services and/or products). In both cases, business risks need to be considered (e.g., the risks of adoption by the organisation of a new technology to give it more advantage in the market or taking risks to drive the market with its own development of new technologies). Other examples include the following:

- Changes in business policy, strategy and objectives;
- Mergers, business downsizing and business expansion;
- Changes in business processes and operational practices;
- Changes regarding the nature, use and application of information and information systems;
- External changes—political and economic, markets, customer base, environmental and social;
- High turnover of human resources;
- Changes to legislation and regulations that the organisation needs to consider to avoid the risk of legal noncompliance;
- New, or changes to existing, contractual obligations, SLAs or contracts that the organisation needs to consider to react to avoid the risk of legal noncompliance or breach of contract;
- Changes to the supply chain driving the organisation or by being driven by the organisation;
- Changes to the technological risks;
- Changes in the overall risk profile due to internal or external events and activities.

Again, in all these examples of change there could be information security requirements, risks and impacts that need to be considered.

8.1.4 Tracking and Reviewing Ongoing Change

The overall ISMS monitoring process is about getting information about all types of change and whether there is the rise in risk and impact levels. Monitoring the result of change may appear through tracking incidents and looking for trends and patterns—whether there has been an improvement or deterioration in the information security performance.

Several processes can be used to provide information about changes and the effects of change. An obvious one is the incident handling process, which provides information about changes to risks and impacts, along with the organisation's ability to respond to these. Reassessing the risks and impacts and doing gap analysis also provides similar information.

Scorecards, staff feedback and similar types of human methods can and do provide some valuable information. Technology in the form of firewalls, IDS devices, audit trail applications, various monitoring tool for networks, services, web browsing and web-based applications can also add considerable value in this information-gathering process, all the time building up a better picture of the changes, the trends and patterns allowing future planning and well-informed management-decision making. Also, the organisation should not ignore the valuable information that could be available from customer and supplier feedback.

8.1.5 Informed Decision Making

Informed decision making in the organisation is essential if the ISMS is to be effective, adequate and suitable to meet business needs. The organisation needs to have a core knowledge of its strengths, weakness, opportunities and threats regarding its ISMS in the context of its business. This includes having a clear knowledge and understanding of its business and legal requirements in relation to its ISMS; the performance requirements in terms of the effectiveness, adequacy and suitability of the ISMS; and the nature and degree of the risks and impacts it faces, its risk criteria and how much risk it is prepared to tolerate and accept. Having a core knowledge and understanding of its strengths, weaknesses, opportunities and threats will enhance informed policy and decision making, and this greatly assists with its ability to manage it ISMS business support and resources.

As mentioned earlier, information security is more of a management, people, process, policy and procedural issue, and implemented and treated this way it is a business enabler to support its business investments and opportunities.

The management of an organisation is responsible for the management of the risks its faces. This includes operational responsibility as well as providing assurance that the security processes themselves are suitable, adequate and effective, including the risk management process, incident handling process, backup process and so on. Risk management in general covers the assessment of the risks, the implementation and deployments of controls and processes by which the risks are managed and maintained at an acceptable level. The system of controls (policy, procedures, processes and so on) deployed by the organisation should depend strictly on the considered risks and impacts. In order that the right security processes, policies, procedures and other controls are suitable, adequate and effective, then informed decision making is essential and critical to the management review process (Section 8.7).

8.2 Monitoring and Operational Reviews

8.2.1 Monitoring

8.2.1.1 Monitoring and Reviewing the ISMS

Monitoring the ISMS is one of the major areas of requirement and activity in ISO/IEC 27001, and it falls under the heading of performance evaluation. There are several aspects of the ISMS an organisation should be considering, including monitoring and review of the following:

- ISMS performance in terms of its effectiveness, adequacy and suitability;
- Changes, risks and impacts and their effect on the ISMS performance;
- ISMS processes and information security controls to check their effectiveness, adequacy and suitability; this includes processes for incident management, risk assessment/treatment, information processing, access control and human resource management;
- Staff awareness, competence and utilisation of the ISMS;
- Efficiency, efficacy and effectiveness of the IT and network services and infrastructure;
- Management of supplier relationships, services, contracts and SLAs;
- Conformance with the organisation's policies and procedures, contractual obligations, laws and regulations.

8.2.1.2 Monitoring and Reviewing Performance

The organisation needs to monitor and review the performance of its ISMS in terms of its:

- Effectiveness: Is ISMS successful in producing a desired or intended result?
- Adequacy: Does the ISMS deliver an acceptable quality and amount of information security?

- Suitability: Is the ISMS right and appropriate for the organisation's needs, purpose and the nature of business?

Monitoring and reviewing the performance of the ISMS helps the organisation to maintain and improve the ISMS. This also means monitoring and reviewing the ISMS in the organisational context (Section 3.1)—ensuring that the ISMS is effective, adequate and suitable in dealing with both internal and external issues and business context and that the ISMS is meeting the needs and expectations (Section 3.2) of interested parties.

8.2.1.3 Monitoring and Reviewing Business Demands and Requirements

It should be clear by now that any activity involving the ISMS needs be done in cognisance and familiarity with business demands and the ability to meet these demands. This should be the true essence and ethos behind the organisation's implementation of its ISMS. Any reviews relating to the ISMS needs to always bear this in mind.

One way of making sure that activities are in sync with such demands and the meeting of requirements of interested parties is to make information security a topic on the agenda of management meetings at all levels, a part of management decision making, a part of the overall business culture and way of thinking and an integrated part of other relevant business activities and process. This is to ensure that those dealing with the ISMS do not lose sight and focus of the business goals and requirements, that senior management is fully aware of the ISMS activities and developments and that continued support and commitment are assured.

8.2.2 Monitoring and Review of Staff Awareness, Competency and Use of the ISMS

8.2.2.1 Competence

Staff competence is generally considered to be the knowledge, skills and behaviour that the organisation requires of an individual to properly perform a specific job function or role. For instance, a sales and marketing management role might require skill in influence and negotiation, whereas a network administrator's role should include technical knowledge and analytical skills. Someone working in the personnel group responsible for all the staff records needs to have a working knowledge of the legislation that applies to the handling of personnel data, information security to protect data, knowledge of handing personnel problems, skill at handling a range of different personnel types, and good communication skills. These are a few of the necessary competence requirements.

It is often said that a part of human nature and psychology tends to make people examine themselves and either under- or over estimate their competence: those that are incompetent at tasks sometime overestimate, while those very competent people sometimes underestimate themselves. Good management should always motivate rather than demotivate their staff, as staff are one of the most important assets the organisation has. Providing good management leadership and motivation, giving the right amount of responsibility and incentives, and enabling good communication channels between management and staff is a pathway to success.

Monitoring staff resource levels, their skill sets, their career development, the need for training/retraining and so on is a major task that management needs to undertake (see Chapter 5).

8.2.2.2 Security Culture

Staff members are significant drivers in their own productivity and the collective productivity and efficiency of the business. Therefore, it important to have good human resources practices in place as well as instilling a good security culture. Developing a healthy security culture within the organisation should be one of the main aims of the organisation. This topic of security culture is so important as a management control for information security that it is being addressed by various international institutions and organisations, including the OECD in its paper "Guidelines for the Security of Information Systems and Networks—Towards a Culture of Security, July 2002."

Good human resource management should be there to deal with internal staff problems or conflicts (e.g., dealing with disgruntled or dissatisfied staff, those operating under poor working conditions or under undue pressure or those that are not being stretched to their levels of skill and competency). Dealing with these issues as soon as possible can avoid some of the insider risks mentioned earlier, whilst supporting staff productivity and efficiency.

Some organisations have had a security culture operating in their organisation for many years. These organisations have realized the importance and benefits that such a culture can offer to the overall awareness and subsequent information security of their businesses as a whole. It is often said that IT is not the problem of information security; rather, it's the people using the IT. This matches up to many of the security surveys, reviews, audits, and reports that indicate that people are the cause of many security breaches and incidents.

8.2.2.3 Monitoring Staff

Monitoring and reviewing staff in relation to their responsibilities and roles regarding information security is extremely important. However, this should be done as part of the normal line management and human resource processes that already exist. It should not be done in a "big brother" way, as this will make staff feel ill at ease, demotivated or overly restricted, which can lead to mistakes or accidents in their work. It might have an impact on their efficiency and efficacy, and it could in some cases lead to a disgruntled employee launching an insider attack. So, a balance needs to be reached between a laissez-faire and a draconian style of human resource management.

Many organisations have adopted an AUP (Section 6.4.2) to control the use of organisational resources and facilities (e.g., IT, Internet services and email).

All forms of monitoring and reviewing of staff activities need to be done in accordance with local laws and regulations.

8.2.2.4 Awareness, Training and Career Development

Developing and improving staff knowledge and skills requires an appropriate level of training and awareness. Formal classroom training, on-the-job training and online training are some of the ways the organisation can help its staff in attaining the right levels of knowledge and skills for the job. As people are of such an importance to an organisation to meet its objectives and performance levels, it is equally important to look after staff training, awareness and career development. Not only does this ensure that the company has the right, competent people to do the job, but it also motivates staff as their career development is being looked after and avoids unnecessary loss of skilled staff that do not receive such attention. Section 8.3 covers example measurements concerning staff awareness and training.

Monitoring staff members regarding their ongoing competence to perform specific job functions enables the organisation to further develop the knowledge, skills and abilities of staff to continue to perform such a job efficiently and effectively whilst protecting the organisation's information.

Of course, the organisation might want to transfer the individual to another job function requiring a different set of competencies and monitoring the individual's ability, knowledge and career development enables the organisation to provide that member of staff with retraining. The human resource group plays a key role in managing the process of matching competence requirements of the job to the skills and knowledge of existing staff or the recruitment of new staff.

8.2.2.5 Some Specific Information Security Roles

The following is an example list of information security roles related to ISMS:

- Chief Information Security Officer (CISO), information security officer(s), manager(s);
- Information security–related management roles and tasks/job functions regarding information systems, human resources, operations, service management, business continuity and so on;
- Information security risk manager(s);
- Information security incident management team;
- Internal ISMS auditor(s);
- Conformance manager(s), PII officer;
- IT- and network-related security roles.

Of course everyone in the organisation will have some part to play in information security, from top/senior management through the various level of management, as well as the nonmanagerial staff.

8.2.3 Monitoring and Review of Information Security Processes

The ISO/IEC 27001 defines many processes, including those at the management, auditing and conformance level, as well as at the human resource level and also at the technical level. These processes will need to be monitored and reviewed to check performance, and if the performance seems to be below standard, then the process needs to be improved. For example, if a backup fails to function properly, does not complete properly, does not deliver according to what is expected, reveals that data recovery is difficult or unsuccessful, generates errors on a regular basis or many other possible problems, then the process needs to be reviewed and improvements to the process are necessary. Other examples of monitoring include the staff recruitment process, the user access management process (granting and removal of accounts, allocation of access rights, privileges and passwords, authentication attributes and so on), service management process or various activities relating to information processing.

The risk management process itself should be reviewed from time to time. The effectiveness of the ISMS is dependent on quality of the risk management: monitoring and reviewing the risk assessment results, the assessment criteria and methods used, the method for determining which controls are to be implemented, the risk treatment decision making, the calculation of the residual risk—all these should be part of the monitoring and review exercise.

8.2.4 Monitoring and Review of Information Security Controls

There are defined in ISO/IEC 27001 many controls designed for different purposes: management, human resource, policies and procedures, auditing, conformance, physical and technical. Again, regular monitoring and review of controls enables the organisation to keep track of its performance and to take appropriate action to implement improvements to the controls as and when necessary. For example, an acceptable use policy may be poorly written, vague and confusing regarding what is and what is not allowable behaviour regarding email usage. Several incidents have occurred whereby users have accidently misused the email system with the root cause seeming to be this poorly written policy. Another example might be monitoring the level of resources needed to deal with the day-to-day management of some of the ISMS processes. If we consider the incident management process as a case in question and the level of team resources needed to do the work: it might be that the number of incidents is steadily increasing or the organisation is experiencing very high bursts of activity. This might demand extra skilled support for the existing incident team either on a full-time basis or a part-time basis, whatever the situation monitoring demand is for ISMS human resource. In some critical cases, this could mean the difference between survival and complete disruption and failure. Monitoring the needs and requirements across all areas of ISMS activity is important.

There are many management controls in ISO/IEC 27001, including controls of the policies and procedures, all of which can be the subject of some form of monitoring. Documentation, such as policies and procedures, will be created, approved, used in day-to-day operations and at some point in time they need to be updated and reviewed. This is normal practice, and part of ISO/IEC 27001 conformance. Monitoring is highly necessary to check that all ISMS-related documentation is suitable and adequate; is available to all those that need to have access; is suitably and effectively protected; has adequate suitable version control; and is adequately and effectively stored, archived and disposed of as is necessary.

Some controls, such as those that are underpinned by the use of IT, may be monitored using automated or semi-automated means (e.g., software used for monitoring unauthorized access or detecting intrusions or for generating audit trails).

8.2.5 Monitoring and Review of IT and Network Services and Infrastructure

ISO/IEC 27001 is not an IT security standard; it is an information security management standard. It deals directly with the security of information, not directly with IT security or network security. Nevertheless, IT and networking services and infrastructure is today a significant management tool and as such, like other management support and business tools, methods and processes, it too has implications on the effectiveness, adequacy and suitability of the organisation's ISMS.

When monitoring and reviewing ISMS use and application of IT and networking services and infrastructure, the following are some of the things that need to be covered:

- User and management policies and procedures that include the secure and acceptable use and application of IT and network resources, user access management, secure information processing involving the use of IT, business communications and transactions using network services and so on;
- Information security management aspects related to the use of business, operational and application software;
- Information security management related to the use of IT for business processes;
- Information security incident management related to the use of IT and networks services and infrastructure;
- Information security management aspects related to the use of IT and networks services and infrastructure for business continuity, availability and resilience.

8.2.6 Monitoring and Reviewing Third Party Contracts and Services

8.2.6.1 Supplier Relationships and Service Management

The majority of organisations have third-party suppliers of a range of different services. Management of supplier relations and service delivery especially considering how services are implemented, operated and maintained also involves consideration of information security.

The organisation should have a policy in place that covers the information security risks regarding the use of suppliers and the protection of the organisation's information.

Also, information security should be addressed in contracts and SLAs with suppliers to ensure that the organisations' requirements are being addressed.

Both the policy and the contract/SLA should address the risks related to the supply IT services and products.

The information security policy, contracts and SLAs should be revised on a regular basis, and of course more immediately if there have been any changes or requests for changes (reduce or expand its current services, revise the current level of service, or add new services). Another reason for revising the contract and SLA is if there have been service-related incidents that would require a reassessment of the risks.

Monitoring and reviewing of third-party services might include checking and evaluating records of service performance, service levels, and any problems in not achieving the requirements of the SLAs, risk levels, service recovery and any incidents that may have occurred.

Clearly the organisation's risk assessment and risk treatment solutions regarding service management needs to be reviewed on a regular basis. In addition, monitoring the levels of information security risk and impact need care and attention to ensure the ISMS still delivers the desired results effectively.

Another aspect is the auditing of the supplier's system to check for compliance against what has been agreed between the contracted parties. The right to audit the supplier's system is something that needs to be agreed between all parties involved—the client and the supplier(s). Such an agreement needs to be included in the service contract, and of course this might stipulate various conditions and restrictions. Some organisations have included this right to audit as part of the tendering process, and some organisations have made third-party certification audits (Chapter 10) a condition of the tender.

8.2.6.2 Monitoring and Reviewing Service Provider Changes

Any changes on the supplier side needs to be communicated to the organisation so they can review whether these will impact them. There will always be a need for IT changes at the service provider end. Maybe their success has secured them more business with existing clients or new clients to meet the increase in demand for IT facilities and services. It might be the case that the provider wants to make changes to stay ahead of its competitors and to be number one in its field of outsourcing, replacing outdated systems or by introducing changes to improve, streamline and/or make more cost effective its operations. Whatever the reason, ICT changes might involve all or some of it hardware, software, operational environments, management tools, networking arrangements, connections and services to change.

Conveying these changes to its clients through regular channels of communications and reporting helps both sides to reassess the risks and plan for any changes that might directly affect the client. This type of reporting should be in the existing contract, and a contract review process needs to be initiated to discuss, negotiate and agree on any amendments or modifications to the contract.

8.2.6.3 Termination of Contract

There are several reasons for terminating a contract: supplier is unable or not willing to expand it services, costs and value for money or lack of information security. It might be that the service provider is taken over or merged with other suppliers, which might present a greater security risk or simply that the service goes out of business, or the supplier has decided to subcontract the work to a supplier located in a high risk area. It might be that the organisation switches suppliers for information security reasons.

Whatever the scenario, there are several information security aspects to be considered. If a termination does go ahead, how is the organisation going to protect its information assets between finishing one contract and starting the next? Does it have the skills in house to protect and manage itself during the transition? The organisation needs to take note that in any transition between suppliers all parties need to be involved to ensure a relatively problem-free changeover, especially to ensure no glitch in the continuation of its operations.

The business continuity and availability issues need to be addressed regarding change in suppliers/providers to ensure that during the transition period there is adequate, effective and suitable protection of organisation's information assets.

8.2.7 Monitoring and Review of Legal and Contractual Compliance

ISO/IEC 27001 specifies that the legal and regulatory requirements pertained to the ISMS need to be taken into account. This also includes contractual requirements and obligations. It is vital that an organisation monitors and reviews and keeps track of its compliance with those laws and regulations that apply to it, reviewing and reassessing the risks and making changes to the ISMS where relevant and appropriate.

With multinational organisations, there is of course potentially more legal and regulatory requirements to keep track of and to comply with. In addition, when trading across the Internet between jurisdictions, there may be several laws and regulations that might apply.

In all cases, the organisation needs to make sure that its ISMS implementation, like all other aspects of operations, should be compliant with all relevant laws and regulations. The impact, for example, in terms of legal penalties, fines, legal punishments for individual company officers, damage to image and so on for failing to comply can be very severe. So knowing what the organisation should comply with and what controls it needs to implement in this respect is part of its conformance with the ISMS specification.

8.3 ISMS Measurements Programme

8.3.1 ISMS Metrics and Measurements

8.3.1.1 Metrics and Measurements

A metric is a standard or a system of measurement, and a measurement is the result of measuring something against a metric. For example, we have the litre as a standard for measuring the capacity of liquid, seconds as a measurement of time or degrees Celsius as a measurement of temperature.

In terms of information security, there are many types of metrics or measures that can be defined (e.g., the number in percentage of system vulnerabilities that are resolved within a defined time after set discovery time, or the number in percentage of incidents that are resolved within a defined time after set discovery time). Another example might be the effectiveness of the security help desk: number (%) of calls received, answered or abandoned; average query processing time and average waiting time; number (%) of resolved/unresolved problems; user/customer satisfaction. Whatever metrics are selected they need to be appropriate to the organisation's requirements to deliver and maintain an effective, adequate and suitable ISMS.

8.3.1.2 ISMS Measurements

In terms of the ISMS, the organisation needs to define what metrics/measures it needs to use, and how, when, where and by whom/what measurements are to be taken. The selection of metrics/measures is entirely in the hands of the organisation, as is the how, when, where and what regarding the measurements. Of course, it must always be focused on the aim of taking measurements (i.e., to check the performance of the ISMS with regard to its effectiveness, adequacy and suitability). These measurements will provide valuable input regarding the monitoring and review of the ISMS and consequently help to determine what improvements need to be made to the ISMS.

Taking measurements and analyzing them is a mandatory requirement of ISO/IEC 27001 if an organisation wishes to claim conformance with this standard. In addition, taking measurements has many business benefits for the organisation (e.g., it greatly assists in managing the risks and the allocation of resource and budget toward controlling the risks). It also helps to demonstrate conformance (i.e., the organisation is meeting the needs and expectations of interested parties). It also helps to increase accountability and to check implementation.

8.3.1.3 Example Areas of Measurements

Measurements can help to indicate the extent to which the ISMS is delivering the desired and expected results (effectiveness), the extent to which the ISMS it is achieving an acceptable level of quality with regard to information security protection (adequacy) and the extent to which the ISMS is right and appropriate for the organisation (suitability).

The following are some example areas to illustrate possible ISMS measurements.

Management Processes

- Effectiveness, adequacy and suitability of the process for the ongoing determination of the requirements, needs, expectations for, and issues related to the ISMS;
- Effectiveness and adequacy of ongoing leadership and support for the ISMS;
- Effectiveness, adequacy and suitability of internal and external communications on all issues, news, changes and improvements related to the ISMS;
- Effectiveness, adequacy and suitability of the ISMS documentation system, including its protection, completeness, currency, availability, distribution, changes and updates.

Human Resources

- Operational performance of staff members and their knowledge and understanding of security procedures;
- Effectiveness, adequacy and suitability of user training and awareness;
- Effectiveness, adequacy and suitability of human resource security—recruitment, during and termination of employment.

Security Processes

- The effectiveness and suitability of the risk assessment and treatment processes;
- The effectiveness and suitability of the information security incident management processes;
- The effectiveness and suitability of the backup processes, vulnerability management processes, configuration and change management process and maintenance processes;
- User access control management processes:
  - User accounts (registration and removal);
  - Managing (granting, changing and removing) rights, passwords and privileges and authentication attributes.
- Effectiveness, adequacy and suitability physical and environmental controls.

Management and User Controls

- Levels of usage and demand for information processing, IT and network services resources to help with capacity planning;
- Effectiveness, adequacy and suitability of policies and procedures:
  - Access control;
  - Acceptable use;
  - Media handling.
- Access methods (to information, applications, systems and networks).

Reviews and Audits

- Effectiveness, adequacy and suitability of in ISMS audits—including specifics such as number of observations and nonconformances found, whether they are major or minor nonconformances and closure rates for nonconformances;
- Gap analyses and reviews of process and control implementation—checking the gaps and weaknesses in the implementation;
- Effectiveness of management reviews.

Economics, Investments and Risk/Impact Costs

- Information security budgets and resources;
- Cost/benefit analyses regarding investing in information security versus impacts;
- Incident management, disaster recovery and business continuity costs; costs of recovery; cost of replacement; legal costs;, contractual penalties; costs against impact analyses and other costs with information security dependencies.

8.3.2 Measurement Programme

8.3.2.1 Programme Design Objectives

An ISMS measurements programme should be designed to provide the organisation with the following:

- A valued assessment as to how well they are doing, how well the ISMS processes and controls are being utilizing and how well they are performing;
- Set benchmarks and performance indicators to be set;
- Better informed decision making about the ISMS effectiveness, adequacy and suitability; risk treatment solutions; making improvements;
- Demonstrate to CEO, directors, senior management, interested parties and the board the value of the ISMS in the sense that risks and impacts are being managed effectively to achieve the desired ISMS results, there is an adequate level of protection provided by the ISMS, the ISMS is suitable to meet the organisation's needs and that it remains so.

Taking measurements should be seen as an activity of great import toward achieving, maintaining and improving ISMS effectiveness, adequacy and suitability. The measurements activity needs to be linked with the monitoring activity, as measurements should be taken on things that are being monitored for effectiveness, adequacy or suitability. Therefore, all that was said in Section 8.2 applies here in Section 8.3. Equally monitoring and measurements are tied into the management reviews, as we will discuss in Section 8.5.

One way to interpret the requirements in Chapter 9 of ISO/IEC 27001 on performance evaluation is to say that the organisation needs to take into account the following:

- A set of metrics (predefined standards of measurements—types of measures) to take measurements against must be defined.
- Part of the task of defining a measurement system will be to define the information security targets and thresholds—key performance indicators and how these targets are calculated, reviewed and updated, bearing in mind that measurements should be objective, reliable, accurate, repeatable, verifiable, relevant and quantifiable against the business targets.
- Which processes, procedures and methods need to be utilized to take and evaluate these measurements.
- The frequency of taking measurements—when measurements should be taken, which staff members are allocated the task of doing measurements and who needs to ensure they are taken (bearing in mind that some measurements could be manual, semi-automated or automated).
- When the measurements need to be checked, assessed, analysed, evaluated and reviewed, and which staff members are allocated this job function.

8.3.2.2 Measurement Methods

The standard ISO/IEC 27001 does not define any measurement methods, as the number of ways of taking measurements is numerous. Also a measurement system is very dependent on the organisation's internal standards, processes, modes of working and business culture. It is likely that the organisation will have performance evaluation exercises in other areas of their business (e.g., project management, quality management or problem management). The methods used in these other areas for performance evaluation might well be adopted for ISMS use. But again this depends entirely on the organisation and its senior management to decide which methods, metrics and measurement systems should be adopted to satisfy the requirements of ISO/IEC 27001.

Some data management methods do apply to the process of measurements and can be used for the gathering and collection of measurement data, and some methods can also help in checking the quality, integrity and validity of the measurement data. But again it should be emphasized that it is up to the organisation to decide and manage its measurements systems and which methods are adequate and suitable for its own use. It should also be emphasized that the monitoring and measurement systems and methods used are themselves a subject for performance evaluation. This means that whatever metrics, methods and the who, when and how of taking measurements also comes under the scrutiny of being assessed and reviewed, as improvements may well be necessary if the monitoring and measurement system that is adopted is itself not performing well (i.e., it is proving to be ineffective, not adequate or unsuitable). So the monitoring and measurement system needs to be reviewed at planned intervals to check its effectiveness, adequacy and suitability.

8.3.2.3 Protection of Data

It is important to note that some of the data being collected as part of the monitoring and measurement exercise may contain details of real or potential weaknesses or vulnerabilities in the organisation's operational systems and processes. As such, this type of sensitive data, in the form of records and other documentation, needs to be adequately protected to avoid leakage, unauthorised access and unauthorised modification.

8.3.2.4 Data Gathering and Collection

What we will present here is some of the basic methods of gathering information for measuring performance. It is frequently said that an organisation only gets what it measures, so if it does not measure the right things, set the most suitable performance targets and indicators or ask the most appropriate questions, it may not arrive at the right answers to tell it how well its ISMS is performing. Of course, this is easier said than done; hence the reason to use continual improvement, which helps to monitor, review and improve processes, including those used for measurements. Therefore, care and attention in the design of these methods and measurements will avoid unnecessary work and provide more appropriate decision-making results.

Additionally, the quality, integrity and validity of data depends on the sample taken (e.g., a poor sample, a small sample or a sample that is not truly representative of a particular aspect of the ISMS). If the results of the measurement programme are to be useful, adequate and suitable to be of benefit to the organisation, then the data needs to be collected from a truly representative, statistically significant, consistent sample.

Another important thing to bear in mind is that the exercise of data gathering/collection and measurements thereof should not interfere with business operations.

As already mentioned, data gathering could be manual, semi-automated or automated.

8.3.2.5 Measurement Templates and Examples

Some organisations use a predefined template for recording measurements. Each type of measurement would have its own predefined template covering a definition of the metric being used, the ISMS objective of the metric, details of the measurement to be taken against this metric, the type of measure (e.g., measure of effectiveness, adequacy and/or suitability), the performance target associated with this measurement, the range and type of data to be collected for this measurement, the frequency of measure, who is the owner and other interested parties for this measurement).

For example, a measure is defined by the organisation regarding the effectiveness, adequacy and suitability of its ISMS training and awareness programme. Goals are to have a staff members with knowledge, working experience and operational competence in the use of ISMS processes and controls; that the training they have received is effective, adequate and suitable for their job functions, their related roles and responsibilities; the measure is a percentage of effectiveness, percentage of adequacy and percentage of suitability as determined by a test, examination and operational demonstration; the percentage targets are set by the organisation; the measurements are carried after each planned training session (e.g., quarterly); the interested parties are human resources, head of training and training managers and line managers.

8.3.2.6 Incident Management and Other Reports

Incident handling reports provide a wealth of data regarding the state of the ISMS to protect the organisation's information. These reports will have details covering what the incident problem is, its nature and characteristics, how and why the incident happened, which systems and process were affected and what was needed to recover from the incident. Incident reports over a period of time provide data on the trends and patterns of incidents, their type and nature. This trend information provides valuable measurement data (e.g., unauthorised access attempts; how many attempts there have been and how many were successful, access to which parts of the organisation and its systems and processes were attempted, types of damage incurred, which systems or processes were compromised and other types of information). All this data and information, of this incident and all other incidents, provide indications as to whether the ISMS provides adequate protection, whether this protection is effective and whether this protection is suitable in the context of the organisation's business and the needs of interested parties. It is therefore essential that the incident management documentation and records is kept, since this system will contain substantial volumes of data that the measurements programme could use.

8.3.2.7 Scorecards and Questionnaires

Scorecards (especially balanced scorecards) provide a means to measure the performance of key business processes against business strategy and objectives. For example, the scorecard could be used to measure the performance of the incident handling processes against the ISMS objectives referred to earlier. This might consider score questions related to the following:

- Incidents reported and resolved over a period time relating to unusual web site behaviour, adware, spyware, phishing and other similar issues across different parts of the organisation;
- How long it took;
- Time spent recovering from system downtimes, failures and/or unavailability of network services;
- Time spent in correcting user errors in data entry and processing activities.

The results of using scorecards provide management with a snapshot of how effective the risk management controls are. It is a tool that can be used alongside other methods that have been mentioned to complement the contribution of information gathering during the monitoring and reviewing process.

Scorecards should be used on a regular or periodic basis to get the most benefit and to support the continuing pursuit of effective information security through the process of continual improvement.

Equally any security questionnaires asking questions about ISMS use and implementation also provides measurement data. If the questionnaire is well designed with targeted questions of particular interest to the measurements programme, then the answers provide another source of useful data.

8.3.2.8 Tests and On-the-Job Exercises

Testing and exams are one method of collecting data to measure performance related to the human resources aspects of ISMS. These are designed to test the knowledge and understanding of staff members with regard to the ISMS they are using. This might be testing knowledge of a set of policies and procedures with example exercises and problems to solve on typical work scenarios. On-the-job training and at-the-desk exercises test the use and application of ISMS processes and controls.

Benchmarking is a useful tool, as it can provide a means of comparing different parts of the business based on a set of metrics aligned to the objectives, needs and expectations of the interested parties involved in the ISMS with regard to the effectiveness of their information security and their level of conformance and progress. It can also be used to evaluate and check how a company is doing with its information security with respect to other companies working in the same line of business or in the same industry sector.

Benchmarks provide an innovative way of viewing information security as a business enabler across different areas of the business identifying areas for improvement, integration, streamlining and greater efficiency and productivity.

8.3.2.9 Gap Analysis and Benchmarking Methods

Gap analysis is another assessment tool that enables an organisation to compare its actual information security profile with what it needs to be compliant with and/or the security objectives and targets it wants to achieve, and to compare the actual performance with its potential performance. The organisation can use this to check against predefined metrics and targets aligned with the organisation's objectives, needs and expectations and that of all the interested parties involved in the ISMS. This provides the organisation with measurements of conformance against set targets (e.g., checking the extent of implementation and operational usage of the standard principles of privacy as set out in ISO/IEC 29100). This gap analysis will provide insight into areas where there is room for improvement.

The gap analysis process provides a means of determining and documenting the variances between the target ISMS business objectives, needs, expectations, requirements, and what is actually implemented in the ISMS. There are many uses of gap analysis:

- Pre–risk assessment to check what controls and processes are currently in place;
- Post–risk assessment, treatment and selection of controls activities and actions;
- Checking status of implementations, actions and status related to management reviews and improvements;
- During ongoing risk management activities such risk reassessments;
- Precertification audits as well as internal audits and reviews;

- Checking operational conformance with ISMS policies and procedures;
- ISMS asset management;
- Protection of information lifecycle management;
- Incident management applications;
- Business impact analyses;
- ISMS resource allocation, provisioning and management.

The granularity of the analysis can vary in several ways for example:

- Targets of conformance complete (C), semicomplete (S), partial (P) or zero (Z);
- The number of questions asked to establish the level of conformance can also vary.

The greater the granularity of analysis, the more precise the results are likely to be. The following example illustrates this idea.

For example, let us assume that the organisation carries out measurements on its service and supplier relationships. The purpose of this is to check that the controls in place to protect the organisation's assets, including its information assets, are effective, adequate and suitable. A single question to assess whether the organisation has an information security policy in place for each of its suppliers would be:

- Has the organisation an information security policy in place to cover the risks associated with access to its information assets by service providers?

This question is very simple and gives no indication of whether policy specifics are complied with. So an expanded set to several questions might be used to give a greater degree of measurement of conformance of the specifics of the policy:

Table 8.1
Example Gap Analysis Template

Question	Conformance Targets				Comment	Action	Closure
	C	S	P	Z			

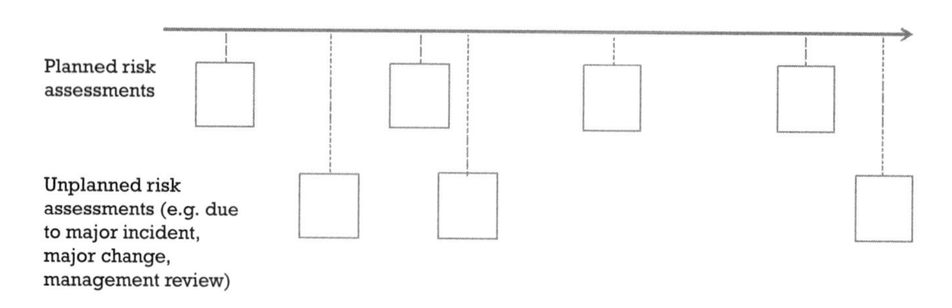

Figure 8.1 Risk review cycles.

- Does the organisation's information security policy for service providers cover controls related to the protection of the confidentiality integrity of organisation's information?
- Does the organisation's information security policy for service providers cover controls related to the protection of all personally identifiable information that the organisation has?
- Does the organisation's information security policy for service providers cover controls related to incident management?
- Does the organisation's information security policy for service providers cover controls to ensure availability of resources used to process information and of information itself?
- Does the organisation's information security policy for service providers cover monitoring the quality of service provided?

8.3.2.10 Automated Data Collection and Processing

Of course, manual data collection, although inevitable for a number of measurement exercises, is not necessarily the most efficient. In order to add greater efficiency the use of automated tools for data collection and processing is far better.

Now there are several ways automation can be used. Clearly, having most of the methods mentioned earlier automated is definitely possible. For example, scorecards, questionnaires and sometimes gap analyses can be done online and so all the results are in electronic form, which makes processing, analysis and future use and application easier.

Other forms of automated data collection relates to reporting of incidents, problems, errors, vulnerabilities and customer feedback; although these may not directly relate to any measurement exercise per se, they are data that is useful in the broader scheme of things and do constitute a measure of the effectiveness, adequacy and suitability of the ISMS.

There are indeed software tools that are specifically designed to collect data, analyse data and present results. IDS software provides indications of traffic and user behaviour against predefined patterns of behaviour; firewalls and other network technologies also provide data related to network usage, behaviour, problems and so on. IT has audit trails built in that provide data on user access to files, applications and services, systems and various other things.

So automated data collection, processing and analysis lends itself to being an integral part of a measurements programme. However, not all of the work that is needed for an ISMS measurement programme can be automated, since much of the ISMS is about people, management, and the use and application of information. Putting automation and IT in context, we can say it is an extremely important business tool designed to make processing of information more efficient, but we cannot totally automate the human aspect, from a usage, behavioural and application perspective.

8.3.2.11 Measurements in Practice

A number of key points should be kept in mind by the organisation in designing an effective measurement programme, such as the following:

- Value and purpose:
    - Provides a measure of performance of the effectiveness, adequacy and suitability of the ISMS against the organisation's objectives and the needs and expectations of all interested parties;
    - Provides an understanding of possible risks (new risks or changes to existing identified risks), and provides an indication that more controls are needed or existing processes and/or controls need improvements;
    - Helps in monitoring the ISMS and provides essential information for the management reviews;
    - Identifies problem areas where improvements could be made;
    - Identifies areas where the performance is below par.
- Type, quantity, quality and timeliness:
    - Make sure the measurements are relevant and a sufficient number of measurements are taken regularly and at the right time,
    - Make sure you are not measuring too many things that are not used—the measurement programme needs to be focused, con-

sider what is relevant, the critical parts and potential risks of the ISMS;
- Make sure imprecise measurements or irrelevant measurements are not being taken.

- Use and interpretation of measurements:
  - Make sure measurements are being undertaken with a clear goal in mind;
  - Make sure all those who are asked to provide input understand the purpose and relevance;
  - Right use (e.g., to gain a better understanding the state of the ISMS).

It is important that whatever metrics are used and whatever measurement system is in place to measure against these metrics will deliver meaningful, useful and timely results. There is always room for improvements, as this is the nature of change and this is why monitoring and measurements play an important role in the lifetime of an ISMS. No one can get a measurement programme fully correct from the word go, and it is unlikely that such a goal in reality can ever be attained because of change. Therefore, the measurement programme itself needs to be reviewed and evaluated, especially the relevance of the metrics being used, the method of taking measurements, the frequency of measurements, and so on.

8.4 Ongoing Risk Management

8.4.1 Risk Responsiveness and Commitment

This chapter is about ISMS performance evaluation, and at the heart of this like all other aspects of the ISMS is the topic of risk—we cannot get away or remove ourselves from this important and critical topic. Measurements might inform us that there is a problem with the ISMS (i.e., there are risks that are causing this problem): new risks, variants of old risks or an old risk that has become more virulent, hostile, more likely or carrying a greater impact. In such a case, the organisation needs to reassess its risks. At the same time, a risk assessment, in and of itself, is a measurement of how good the ISMS is in terms of effectiveness, adequacy and suitability.

Taking suitable action and being responsive to manage the information security risks an organisation faces is crucial to achieving an effective, adequate and suitable ISMS. The processes of risk assessment, risk treatment and selection of controls constitute the first phase of the risk control procedure (as discussed in Chapter 4). The second phase encompasses implementing the controls and supporting actions to put in place a system of controls and getting them deployed in the ISMS (as discussed in Chapters 5–7), and the third phase involves keeping ahead, being present with the latest changes, being responsive to changes where appropriate and maintaining an effective information security regime (the topic of this chapter).

At all stages of the risk management process, there needs to be a firm commitment from senior management, operational managers and staff. All have a part to play to counter the threats and incidents that result in unwanted business risks and impacts. Table 8.2 shows a simplified view of typical levels of commitment and responsibilities.

8.4.2 Regular Risk Assessments

From time to time, the organisation's risk profile needs to be reviewed and updated. This is essential to keep up with changes, to make improvements and to maintain an effective, adequate and suitable ISMS implementation. It is a key part of the continual improvement process (Chapter 9).

This means that risk assessments should be carried out on a regular basis. In addition, assessing the risk will also be needed in the case of a major incident or change in the business. This will include reviewing the current status of previously identified risks and impacts as well as considering new

Table 8.2
Levels of Commitment

Human Resource	Level of Commitment and Example Responsibilities.
CEO, directors, senior management	Commit funding, recruitment of staff and other resources for the management of information security risks, ISMS establishment, implementation, monitoring, reviews and measurements, and continual improvement. Set corporate policy, strategy and business objectives for risk management activities. Establish risk criteria. Ultimately accountable for the management and control of the risks and impacts. Establishing and visibly support an information security culture, including training and awareness. Define and allocated roles and responsibilities.
Operational management and other areas of middle management activities (such as human resources, administration)	Implementing policy and business objectives in the form of operational risk management policies and procedures. Measuring operational performance, effectiveness and efficiency. Management of operational staff and their deployment of risk controls. Recruiting staff. Setting operational risk control responsibilities providing appropriate staff awareness, training and on-the-job training where necessary. Monitor, review and report back to senior management risk-related issues, incidents and risk management performance.
Staff and users	Being aware and responsible for information security risk management in their specific areas of work, and understanding information security as an integral part of the work. Reporting of security incidents and feedback on effectiveness of security procedures. Being responsive and working as part a collective team throughout the organisation to manage the risks.

risks and impacts. The risk register will then need to be updated. As an example this would typically include the following:

- Previously identified risks and impacts:
    - Review, reassess and recalculate previously identified risks and impacts, checking levels and status appropriateness,
    - Add, modify and/or delete entries or details of all relevant changes in the risk register.
- Newly identified risks and impacts:
    - Assess newly identified risks, taking account information contained in incident reports; audit reports; security reviews; customer, user and management feedback and reports; any changed conditions and any new risk relevant information;
    - Calculate the impacts related to the new risks.
- Reassess the risk priorities.

8.4.3 Risk Measurements and Metrics

Different types of metrics and measurements are available that relate to the risk management–based process.

Quantitative metrics and measurements are based on the use of statistical or historical data and monetary values. This includes assets where real quantities can be used to associate damage directly with some numerical value:

- Cost of recovering from a system failure, virus attack, denial of service and other incidents;
- Cost of replacing a damaged system or piece of equipment;
- Reliability figures of equipment or component failures (e.g., MTBF);
- Records relating to system down times, access attempts, network data (e.g., transmission delays, access delays, call and session drops and error rates);
- Statistics relating to earthquakes, tsunamis, volcano eruptions, flooding, typhoons and other natural threats, pandemics and so on.

Some assets are difficult to assign a quantitative risk measure to (e.g., information itself, company image and reputation, the skills of human resources or the competence of the attacker and their capacity and resources for carrying out the attack). Therefore, quantitative methods for ISMS risks and measurement are not always possible to apply due to lack the data available, in which case qualitative methods are normally applied.

Qualitative metrics and measurements do not involve the use of statistical or historical data but use a grading system such as low, medium, high and very high to express the value of the risk, impact or likelihood. Although this may seem an imprecise method, it is generally a good method to deal with the broad range of objects being dealt with in ISMS—people, management, processes and information. Hence, qualitative methods are generally more suited to an ISMS since there is often more uncertainty in assessing the information security environment. In practice, the complexity of information security issues requires a combination of qualitative and quantitative assessments, especially as the assets will cover tangible and nontangible, management and nonmanagement asset types.

Semi-quantitative methods address the need to express qualitative measurements in the form of numbers (e.g., low risk = 0, medium risk = 1, high risk = 2 and very high risk = 3). The numbers themselves can be further quantified in more meaningful terms, such as monetary values, degradation in levels of service and availability, reduction in performance levels and decrease in efficiency and productivity output.

8.5 ISMS Internal Audits

To comply with ISO/IEC 27001, organisations need to conduct internal audits. The purpose of an internal audit is to check whether the ISMS satisfies the requirements of ISO/IEC 27001. This will consider the effectiveness, adequacy and suitability of the ISMS. This audit might identify weak links and gaps in the ISMS and potential opportunities for improvements.

These audits should be carried out at planned intervals as part of an audit programme. Each audit should have an audit plan that details the objectives and scope of the audit, the audit criteria. The objectives of the audit is determining what needs to be achieved by the audit, including conformity of the ISMS with the audit criteria, and determining whether the ISMS meets the relevant legislative, regulatory or contractual requirements. The audit scope will define the extent and limits of the audit—the parts of the ISMS that are to be covered: which activities, systems, processes, controls, locations, business units and so on. The audit criteria shall include the requirements in ISO/IEC 27001 and any standards or requirements specific to the organisation.

The audit shall be carried out by a qualified ISMS auditor or team of ISMS auditors, depending on the size and extent of the audit scope. A qualified ISMS auditor (or auditor team) will have the necessary knowledge, skills, competence and audit experience to undertake ISMS audits. If there is an audit team, then the team will be headed by a lead auditor, and typically the lead auditor should have several more years of experience in ISMS auditing. Those that carry our ISMS audits should not have been involved in the development of the ISMS or have any current involvement in ISMS activities. In other words, they need to avoid any conflicts of interest and they need to remain impartial. It is only through the impartiality of the audit that confidence and trust in the results can be assured. Internal ISMS audits, and third-party certification audits, therefore need to be objective, unbiased, free of prejudice, balanced and fair.

The general audit process, which is applicable to internal audits, is described in Chapter 10. Also, more details and guidance on audits can be found in ISO 19011, ISO 17021 and ISO/IEC 27007.

The results and findings of the internal audit, documented in an audit report, are provided as input to the management review meetings (Section 8.6). The results also provide feedback to the senior management and interested parties.

Internal audits are a key element in the monitoring and review process, and they are a useful management tool for identifying problems, risks and nonconformities related to the ISMS.

8.6 Management Reviews of the ISMS

8.6.1 Management Review

Another aspect of the performance evaluation exercise is management review of the ISMS. Senior management shall have management reviews regularly at planned intervals. The purpose of these, like that of the overall aim of performance evaluation, is to check the effectiveness, adequacy and suitability of the ISMS. These meetings need to review all relevant information: reports from audits and security assessments and results of measurements, risk assessments, feedback, changes affecting the organisation, changes to requirements and consideration of other types of input that might have an impact on the effectiveness, adequacy and suitability of the ISMS.

These meetings should be attended by interested parties and all those who need to report into such meetings, because of their ISMS job function or because of an action that was placed on them (e.g., by past meetings). Others might need to attend these meetings, not on a regular basis but where a specific agenda item requires their specific skills or knowledge about a relevant ISMS subject.

The agenda and minutes of these meetings shall be documented: this should be a record of discussions, decisions, actions and recommendations for ISMS improvements.

8.6.2 Input for the Management Review

Gathering information concerning changes should be part of an organisation's monitoring, review and reporting process. The organisation can get information, for example, from the following

8.6.2.1 Incident Handling Records

These records can provide information on the organisation's ability to resist against and respond to threats to its information systems. These records are a valuable source of information regarding the effectiveness of the organisation's ISMS. Regular reviews of this information is essential to keeping up to date regarding the organisation's responsiveness and preparedness for threats and attacks. Incident handling is key to the ISO/IEC 27001 ISMS process.

8.6.2.2 ISMS Measurements

Setting up metrics, a measurement programme and performance indicators is key to the ISO/IEC 27001 ISMS process. How effective, adequate and suitable are the leadership and support controls that management have in place? Is the training and awareness programme effective? How effective is the management of human resources? How effective is the incident handling process? How well does IT backup perform? How good is the performance of the access control system to withstand attacks from external networks? These and many other questions can lead us to judge how effective our ISMS is at managing risks based on what we measure.

8.6.2.3 Scorecards, Gap Analysis and Benchmarking Processes, Tools and Exercises

These methods, if used on a periodic basis across the organisation and by an appropriate number of staff, can provide a valuable source of information. They give a relatively easy way of determining whether security is being deployed, whether it is working and whether target levels are being met.

8.6.2.4 Internal Audits and Reviews and Third-Party Certification Audits

All forms of audits and reviews are important to determine if the ISMS is conforming to the requirements of ISO/IEC 27001, and whether the system of controls to manage the risks have been properly implemented, deployed and maintained. It is also necessary to review other audit results and to provide reports to management on the status of information security within the organisation.

8.6.2.5 Feedback from Users, Employees, Customers and Suppliers

Often valuable information on the effectiveness of the organisation can be gotten from staff. Of course, this should be the case since they are in daily contact with the system of risk controls in place. They are using the procedures that underpin the management and application of the information security measures in place. Also, customers and suppliers can sometimes provide useful information, both positive and negative, about the effectiveness of the organisation's measures to protect information.

8.6.3 Output of the Management Review

8.6.3.1 Recommendations for Improvement

Gathering information and evaluating and analyzing it with respect to the effectiveness, adequacy and suitability of the organisation's ISMS implementation is a key exercise and one that proves essential input into management review. The next phase is that of deciding whether improvements need to be made, based on this input. What is decided—the recommendations and the actions to be taken—are the main outputs of the management review.

Any changes that are recommended should be to improve the ISMS—to change the ISMS for the better. Of course, any changes to the ISMS to make improvements will have an influence and impact on the business. So any recommendations need to be carefully considered, for although it is aimed at improving the ISMS it may have other impacts on other parts of the organisation's systems and processes.

Recommendations may be made requiring corrective actions to resolve existing problems and to make improvements. It is also the opportunity to take preemptive actions (to implement preventative controls) to prevent further problems from occurring, either old problems reccurring or the anticipation of new problems occurring.

8.6.3.2 Action Plan

Any recommendations and actions to implement ISMS corrective and preventative actions to make ISMS improvements shall have full management commitment and support. An action plan needs to be produced with a prioritisation of the work. After implementing the action plan, it needs to be confirmed that the new improvements are working effectively and are being deployed correctly in the working environment. Reworking, revising and updating the policies and procedures may need to be instituted. The improvement phase might need to confirm whether certain processes need to be improved and modified accordingly.

It is not only ISMS controls that may need to be improved, but also the ISMS processes themselves that may need to be improved and overhauled—the risk assessment and treatment process, the measurements programme and so on. This might entail changes to risk criteria, different methods of risk assessment deployed or new measures and performance indicators may need to be defined or existing ones modified.

8.7 Awareness and Communications

Key to any action senior management might need to take, in this case with regard to ISMS improvements, is effective communication with all managers and staff and all interested parties: awareness and training and deployment of procedures. Users and employees need to be aware of the changes to implement the improvements and how these relate to their own job functions and the work responsibilities they have been assigned. Indeed, they need to practice deploying the procedures and processes they need to follow regarding the management of risk in their areas of work.

Retraining may need to take place if the improvements involve totally new controls or upgrades to existing controls, in particular with those staff members that are directly impacted by these changes.

Communications need to be two-way. Management sets the policy and strategy, makes decisions and provides risk management resources. Users and staff on the other hand should be reporting back to management incidents, giving feedback and providing information on potential risk situations and possible improvements to management; this is valuable input for the organisation itself and for making informed decisions on how best to manage risks and take the most appropriate course of action.

CHAPTER

Contents

Improvements to the ISMS

9.1 Continual Improvement

9.1.1 Improvement

The process of improvement is referred to as a continual improvement in ISO/IEC 27001:2013 and is a mandatory aspect of conformance with this standard. ISMS continual improvement is a top management commitment that should be driven by the objectives set by top management to have an ISMS that is effective, adequate and suitable for its business purposes. This means to achieve continual improvement the organisation needs to consider the issues and requirements defined in ISO/IEC 27001:2013. This means to ensure that appropriate protection is in place to treat the risks—this could be information security controls and processes to detect, prevent, reduce and respond to undesired effects (e.g., from security incidents, breaches, disasters, compromises). It also implies that the ISMS produce the results and outputs that are expected.

The top management of the organisation should always be looking for opportunities for continual improvement. One aspects of this is through the organisation's performance evaluation activities (e.g., from management review meetings Clause 9.3 [ISO/IEC 27001:2013]). It might be the case that the organisation has received feedback from interested parties, such as its business partners or customers, on service delivery and availability. For example, business partners or customers might now be requesting information security requirements to be taken into account in addition to that already determined through a previous consideration of Clause 4.2 (ISO/IEC 27001:2013). It might be that there are complaints or concerns from business partners about a recent spate of incidents the organisation was involved and they want to some reassurance that the organisation is doing something to improve its information security. For example, an organisation experiences an increase in the number of hours of operational down time and subsequently the unavailability of access to information for business processes. This problem then had an impact on business efficiency and customer dissatisfaction. After incident and problem analysis, it appears that the down time is due to the lack of availability of a number of business processes that rely on IT systems. The cause of the problem is due to a lack of regular maintenance, which is subsequently due to a lack of human resources and prioritisation of jobs (i.e., maintenance of critical business systems). Improvements are clearly needed in this case, and action needs to be taken before the problem gets out of hand and the business disruption gets to a point where the impact is highly damaging to the organisation's productivity, efficiency and reputation (in the case where the problem has an effect external customers).

It could be that during a management review the results of the monitoring and measurement activities has highlighted a number of areas where there is opportunity to improve the performance of the ISMS. This could equally be the case with the results of an internal audit or from external third party audits.

Therefore, maintaining and improving the effectiveness, suitability and adequacy of the ISMS is key to ensuring an appropriate level of protection of the organisation's information. Its information policy and objectives are being conformed to be compatible with its business direction and strategy.

9.1.2 Maintaining Effectiveness, Suitability and Adequacy

If an organisation does nothing (i.e., does not make ISMS improvements), then the protection of its information will over a period of time loses its effectiveness. This may happen over a period of months, weeks or even days, and the longer the problem is left unattended, the greater the impact. As discussed in Chapter 9, changes occur—some are in the organisation's control and some not in its control. The degree of uncertainty regarding information security risks in today's fast changing business environment means that organisations need to be proactive and adaptive to the risks brought about by such changes. Improvements will need to be made to ensure the ISMS remains effective in the light of such changes and subsequent risks the organisation faces. The information security risks that organisations are being confronted with are growing in frequency and complexity. Because of the uncertainty of the risks involved, these might be caused by changes that are occurring monthly, weekly or daily in some cases. Improvements are necessary to keep up date with the changes to its risk profile. Some changes may not be classified as information security risks per se but nevertheless can have an impact on the organisation and the effectiveness of its ISMS to protect its information. These changes could be driven by external factors (e.g., changes to suppliers, environmental, infrastructure, conditions, legislation or market conditions). The changes could also be driven by internal risk factors (e.g., changes in business policy, strategy, staff resources, mergers, management decisions, or changes to technology).

Continual improvement is an essential element in the ISMS lifecycle through all the lifecycle stages of implementation, organisational and operational deployment, monitoring, measurements, reviews, maintenance, keeping up to date and improving the ISMS to achieving effective information security management.

9.1.3 Holistic Effectiveness

Continual improvement of the ISMS does not just mean improvement of the effectiveness of the information security controls or processes. The concept applies to the overall ISMS itself. Continual improvement is also a recurring step-by-step activity of improvement and not a continuous process of improvement. This is important since the organisation needs to achieve the right balance between identifying when improvements are necessary and when they should be implemented (based on available resources and level on risk and business impact) as opposed to continually making improvements hour by hour, day by day. Clearly, if risks and impacts have been identified that could compromise the organisation and it ability to protect its information, then it needs to review its situation and take whatever action is most appropriate to manage the risk. Of course, if there has been a major incident, business crisis or emergency, which could have an immediate impact on the well being or survivability of the business, then action needs to be taken that may or may not involve ISMS improvements.

9.2 Conformance and Nonconformance

9.2.1 Nonconformity

An ISMS nonconformity is a nonfulfillment of a requirement of ISO/IEC 27001:2013. This means a nonfulfillment of any of the "shall" requirement in ISO/IEC 27001:2013 is a nonconformity. For example, there is a requirement that management shall review the organisation's ISMS at planned intervals. In fact, all the requirements in ISO/IEC 27001 include the word "shall." The organisation shall do risk assessments according to a planned schedule, it shall undertake a risk treatment process to select the appropriate options for treating the risks (managing the risks), it shall determine an appropriate set of controls and so on. (For the exact wording of these statements the reader, is referred to the text in the standard itself.) One of the tasks of an audit (internal or third party) is to check whether there are any nonconformities, and if nonconformities are found then the organisation should do whatever is necessary to remove such nonconformities.

9.2.2 Corrections

Once a nonconformity has been detected or reported, then a correction can be implemented to eliminate the nonconformity or a corrective action can be taken to eliminate the cause of the nonconformity. For example, if a current policy does not conform to a requirement of the standard, then this problem could be corrected by replacing this policy with one that does conform. Likewise, if a staff awareness programme is insufficient and so results in a nonconformity, then the programme should be replaced with an improved programme.

9.2.3 Corrective Actions and Root Causes

In the case of corrective action, it is necessary to first determine the cause of the nonconformity. Determining this cause depends on the complexity of the problem. In some cases a simple brainstorming exercise is all that is needed; in other cases, the root cause may be found through considering several different contributing factors (a chain of cause and effects), and a more detailed analysis might be needed. The root cause might be a people, process, method, management or technical problem or a combination. The effectiveness of the corrective action will of course depend on whether the cause (or causes) have been correctly identified. The cause may have been accidental (a one-off mistake or error), or it might have been intentional or systematic, which could be a recurring problem if not dealt with. If the cause is not properly identified or analyzed, then a systematic cause may treated as if it were a one-off problem or accident, and then there is a likelihood of the problem recurring.

After determining the cause(s) of the nonconformity, corrective action needs to be taken to eliminate the cause of the nonconformity. In general, then, the organisation's response to a nonconformity should therefore be correction, cause analysis and corrective action or alternatively cause analysis, correction and corrective action. But merely correcting an existing nonconformity may not wholly resolve the problem—the root cause—and so a recurrence of the problem may happen.

9.2.4 Some Common Causes of Nonconformity

There is no definitive list of causes of nonconformities. The following are some examples of nonconformities:

- Management failure to understand the internal and external issues and the requirements of interested parties that are relevant to the ISMS, which subsequently may have an impact on delivering the expected outcomes of the ISMS in conformance with ISO/IEC 27001 Clause 4.1 and 4.2;
- Failure to adequately define the ISMS scope in conformance with ISO/IEC 27001 Clause 4.3—especially a failure to properly address the interfaces and dependencies;
- Lack of an adequate or effective risk assessment and treatment process in conformance with ISO/IEC 27001 Clause 6;
- Management failure to commit to or to achieve continual improvement of the ISMS ISO/IEC 27001 Clauses 5.2 and 6.1.1;
- Lack of resources to support the ISMS in conformance with ISO/IEC 27001 Clause 7;
- Lack of training and awareness in conformance with ISO/IEC 27001 Clause 7.3;
- Failure to implement and control the processes necessary to effectively and adequately address the risks and opportunities in conformance with ISO/IEC 27001 Clause 6;
- Management failure to adequately address the requirements for performance evaluation in conformance with ISO/IEC 27001 Clause 9, including failing to comply with internal audit results, lack of commitment to management reviews and failure to determine a suitable

programme of monitoring and measurements, all of which can fail to identify opportunities for ISMS improvements.

9.2.5 Case Study One

An incident occurs regarding the operational use of security procedures to achieve the organisation's security objectives. The incident is a failure to protect commercial information. A new inexperienced member of staff uses an "acceptable use" policy and procedure that was determined as necessary from a past risk assessment and risk treatment activity (ISO/IEC 27001 6.1.2 and 6.1.3). The inexperienced member of staff sends sensitive commercial information unprotected in an email based on their interpretation of the policy. The initial reaction was that the use of the procedure by the inexperienced member of staff was the cause, but after reviewing the procedures it was found that the instructions given on this issue were unclear—in fact, poorly written. So improvements should be made involving revision of the procedures and, in addition, providing better training to new recruits (ISO/IEC 27001 Clause 7.3).

9.2.6 Case Study Two

An auditor raises a nonconformity regarding management not providing the appropriate level of resource for the maintenance of the ISMS (ISO/IEC 27001 Clause 7.1), in particular measurements for the evaluation of the performance of the ISMS. This results in a lack of appropriate input to the management review meetings (ISO/IEC 27001 Clause 9.3.c). There is a lack of human resources to undertake performance measurements and to analyse and evaluate the ISMS effectiveness as the cause of the problem. Therefore, management needs to review this resourcing problem to improve the process of providing input on the performance of the ISMS to management review meetings.

9.2.7 Case Study Three

An incident occurs related to the organisation's use of social media as part of its marketing campaign. In the risk assessment the organisation had identified, analysed and evaluated a number of risks (ISO/IEC 27001 Clause 6.1.2) relating to its use of social media for business purposes, and it subsequently determined a number of controls (ISO/IEC 27001 Clause 6.1.3b) to treat and manage these risks. Unfortunately, after six months, the organisation has still only implemented a few of these controls, leaving the organisation wide open to the risks assessed. The organisation, however, had failed to implement the controls (ISO/IEC 27001 Clause 8.1) determined in (ISO/IEC 27001 Clause 6.1.3b) to achieve the organisation's security objectives (ISO/IEC 27001 Clause 6.2). The lack of implementation of appropriate risk-based controls was the main cause of the incident (i.e., lack of controls that need to be implemented to treat the identified risks).

9.3 Making Improvements

9.3.1 Planning and Implementing Improvements

Information on what improvements are needed, when improvements should/can be carried out, and what resources are needed should all be documented as part of an improvement plan. The improvement process should ensure that:

- Improvements that are implemented are communicated to all interested parties; in particular, communicate to all those who need to know the details about the improvements (at a level of detail that is appropriate and relevant to them and their job function), whether it is regarding new security methods and measures, new/revised policies and procedures or other changes. Communications might not be just to internal staff but also might need to be to third parties, customers and other external parties who need to know about the changes (i.e., at a "need-to-know" level of detail that is appropriate and relevant to them in accordance with the contracts and SLAs in place) (see ISO/IEC 27001 Clause 7.4).
- Action is taken to revise and update policies and procedures where necessary and appropriate as part of achieving continual improvement.
- Action is taken to revise processes where necessary and appropriate as part of achieving continual improvement (see ISO/IEC 27001 Clause 8.1).
- Suitable awareness is undertaken to inform about the improvements. This might involve changes/revisions to policies and procedures, other risk controls or processes, and it might also involve giving some staff training (see ISO/IEC 27001 Clauses 7.2 and 7.3).
- There are adequate resources available to deploy and utilize the implemented improvements (see ISO/IEC 27001 Clause 7.1).

9.3.2 Improvements to Processes

There are various situations where processes need to be improved (e.g., where processes are not properly used or not followed, are difficult to use, are not well designed or implemented), and they can cause performance problems or result in nonconformities. It might be the case that the risk assessment process itself needs to be improved, as it is not producing consistent or valid results. Another example might be the process that the organisation uses to communicate with external parties needs to be improved. Also, the processes that the organisation deploys for outsourcing are not properly or inadequately controlled. The internal audit process may need to be reviewed and improved to ensure that the impartially of audits are being maintained. There are many processes, both information security processes and business processes, where there could be opportunities for improvement.

9.3.3 Improvements to Policies and Procedures

There are various situations where policies and procedures need to be improved. Policies and procedures may be written in ways that are confusing or difficult to use; they may contain errors, mistakes or provide wrong information; they may be incomplete or not cover all possible situations; they might need updating to take into account new developments. All these factors can and do cause problems in achieving the organisation's security objectives and the intended product of the ISMS. Is the information security policy appropriate to the organisation's objectives and strategic mission? Does the policy cover the right things? Has the organisation a suitable set of procedures necessary to implement the information security objectives? Are the procedures sufficiently clear and understandable to avoid any errors or mistakes when used? Does the scope of the procedures need to be extended and the content updated to cover new areas of development and change in business operations?

9.3.4 Implementing Improvements to Awareness and Training

Lack of training as mentioned earlier is a common cause of a nonconformity and is always an area where there is opportunity for improvement. Does the organisation need to improve the competence of some staff involved in ISMS-related activities through appropriate education and/or on-the-job training and experience? Do staff members need refresher training as relevant to their work? Is the organisation's awareness and training programme sufficiently comprehensive to cover the scope of its ISMS implementation?

CHAPTER

10

Contents

Accredited ISMS Certification

10.1 Overview

This chapter explores the world of third-party ISMS certification: the basic ideas, which parties are involved and what their roles are, what the process involves, who does the audit and what qualifications are needed to be an auditor and finally the current international trend for ISMS certification.

10.2 International Certification

10.2.1 Global Take Up

ISO/IEC 27001 ISMS certifications are carried out in more than 70 countries involving organisations small, medium and large, from a diverse range of business sectors. Some examples are as follows:

- Telecoms and network services and suppliers;
- Financial institutions and insurance sector;

- Manufacturing sector,
- Utilities (electricity, gas and water);
- IT vendors, suppliers and services;
- Government departments and agencies;
- Retail sector;
- Entertainment industry;
- Research and development;
- Professions (e.g., legal profession);
- Academic sector.

The current growth trend is Asia, followed by Europe, then the Americas and finally Africa and the Middle East. The current breakdown (at the time of this writing) is as follows:

- Asia (including Australia and New Zealand): 47%;
- Europe: 29%;
- Americas: 15%;
- Africa and the Middle East: 9%.

Even though this percentage looks asymmetric, the growth rate in all areas of the world is continually increasing, and predictions from the market are that the near future will present a more balanced global spread.

10.2.2 Motivation

This diversity not only demonstrates the importance of information security in every business sector, but also the wide applicability of the ISO/IEC 27000 series of standards to meet the demands of business for a "common language" for addressing information security issues. The motivation for going for a third-party ISMS audit is also varied; for example:

- To comply with legislation and regulations (e.g., data privacy and protection, governance and computer misuse and hacking);
- As part of customer and supplier chain contracts (more tenders and RFPs are now including ISO 27001 compliance clauses);

- To demonstrate "fitness for purpose" (see Chapter 2);
- Insurance reasons (there are instances where insurance premiums have been lowered as organisations can demonstrate they have an effective risk management system in place);
- Market competition (demonstrating ISMS excellence in a particular sector).

10.2.3 Costs and Resources

The cost to the end user organisation of implementing an ISMS and having ISMS certification breaks down into components: the cost of the certification bodies (CBs) certification work (initial audit, follow-up surveillance audits and recertification audits) and the cost of internal work (preparing for certification, ongoing surveillance audits and maintenance and operational costs). How these costs are calculated varies and is dependent on several factors (i.e., the complexity of the ISMS being implemented). In ISO/IEC 27006 Annex A.1 and A.3, there are some examples of what makes up this complexity as well as some guidance on calculating auditor time. With regard to the internal costs, this includes the following:

- Reviewing risks and taking measurements of ISMS effectiveness;
- Implementing and deploy ISMS controls;
- Producing documentation;
- Awareness and training;
- Resources for specific security functions (e.g., a security officer/manager, firewall administrator or physical security manager). Of course, this might seem an expensive exercise, but it should be realised that some of these items should be everyday expenses for a lot of companies since documentation, training and awareness are required in other areas of the business. Also, depending on the size and system's complexity, these costs maybe not be large at all. In effect, if security is integrated, as it should be, with other parts of the business, then some of these costs will be absorbed and shared by other business functions and operations. If information security is seen as an important aspect of the business, and it is rare to find a company where some aspect of information security is not required, then resources will need to be part of the business budget regardless of whether the organisation goes

for certification. Of course, it is a commercial balance and a tradeoff between implementing just the appropriate level of security and going over the top with unnecessary expenditure. This is why it is important to carry out an information security risk assessment—so that management can make the right decisions on where to spend money (see ISO/IEC 27001 Clause 6). Hence, we must not lose sight of the fact that the business needs to survive in today's dynamic and competitive markets, and the more accurate the analysis of the risks the company faces, the better informed management is to make the right decisions to do the best for the company and how best to invest in information security.

10.3 Certification and Accreditation

10.3.1 Interested Parties

Accredited ISMS certification involves the following interested parties: accreditation bodies (ABs), certification bodies (CBs) and the end user organisation (whose ISMS is to be certified). Each of these interested parties play different roles in the process and has different responsibilities.

10.3.2 Accreditation

Accreditation is a formal process carried out by a national accreditation body (AB). The AB is the body that provides formal, third-party recognition of the competence of the CB to perform specific certification and auditing tasks and deliver certification services to the end user. For the end user organisation (CB customer), it provides a means to identify a proven, competent CB so that the selection is an informed choice. It also means the CB can demonstrate to its customers that it has been successful at conforming to the requirements specified in international accreditation standards for management systems.

ISMS certification process uses the same model as other management systems, such ISO 9001 for quality management, ISO 14001 for environment management and ISO 220002 for food safety management.

In assessing the adequacy of the CB to carry out ISMS certifications in accordance with ISO 17021 and ISO/IEC 27006, the AB will:

- Carry out CB head office visits to check that the CB has the necessary systems processes, and procedures in place and these are functioning correctly to undertake such ISMS audits;

- Assess whether the CB has an adequate number of qualified auditors to carry out ISMS certifications;
- Carrying ISMS onsite witnessed audits to assess the competency of the auditors working in practice.

After being accredited, the CB can then offer ISMS certification services to their clients. This accreditation lasts for three years, during which their competence and performance will be continually monitored and assessed by the AB through regular witnessed audits. After the three years, the CB can be reaccredited by the AB, extending its certificate of accreditation for another three years.

The International Accreditation Forum (IAF) site (www.iaf.nu) provides a list of recognised accreditation bodies, and www.european-accreditation.org gives a list of the European accreditation bodies. Examples of ABs are UKAS in the UK, SWEDAC in Sweden, ANAB in the US, EMA in Mexico, TGA in Germany, SCC in Canada and JAB and JIPDEC in Japan.

A CB that that offers accredited certification services need to be after accredited by an AB. Currently, a large number of CBs can provide accredited ISMS certification services.

10.3.3 Certification

The CB carries out certification audits on their clients ISMS. In this chapter, clause 11.5 describes the audit process, which is a two-stage process.

The CB issues certificates to it clients based on a successful audit being carried out. This certificate lasts for three years. During this three-year period, the CB will carry out regular surveillance audits, every 6 to 12 months. At the end of the three years, the certificate expires. If the client wishes to continue to have a valid certification, then before the expiry date the client needs to apply to the CB for a recertification audit. The client will be reissued with a certificate following a successful recertification audit.

10.4 Standards Involved

The main standards involved in the accreditation and certification process are outlines in the following sections.

10.4.1 Accreditation

The following are the current standards used by the AB to assess and accredit the CB:

- ISO 17021-1: Conformity Assessment—Requirements for Bodies Providing Audit and Certification of Management Systems;
- ISO/IEC 27006: International Accreditation Guidelines for the Accreditation of Bodies Operating Certification/Registration of Information Security Management Systems;
- ISO 19011: Guidelines for Auditing Management Systems.

10.4.2 Certification

The following are the current standards used by the CB to audit and certify the customer's ISMS:

- ISO/IEC 27001: Information Security Management Systems—Requirements;
- ISO 19011: Guidelines for Auditing Management Systems;
- ISO/IEC 27007: Guidelines for Information Security Management Systems Auditing;
- ISO/IEC 27008: Guidelines for Auditors on ISMS Controls.

10.4.3 End-User Organisations (ISMS Owners)

Organisations of course may use all the other standards in the ISO/IEC 27000 ISMS series as supporting guidance and advice on the implementation of ISO/IEC 27001. This includes ISO/IEC 27002—Code of Practice for Information Security Controls; ISO/IEC 27003—Information Security Management System—Guidance; ISO/IEC 27004—Information Security Management—Monitoring, Measurement, Analysis and Evaluation; 27005—Information Security Risk Management. In addition, those organisations working in certain business sectors, such as telecom or energy, may decide to use the relevant sector-specific standard for additional controls, and implementation advice—for example, ITU X.1051 | ISO/IEC 27011 (telecoms), ISO/IEC 27013 (service management and ISMS), ISO/IEC 27015 (finance sector), ISO/IEC 27017 and 27018 (cloud services) and ISO/IEC 27019 (energy sector). However, it needs to be emphasised that conformance (in ISO terms) is to the requirements of ISO/IEC 27001 and all other standards are not mandatory.

10.5 ISMS Audits

10.5.1 Certification Scope

The scope of the ISMS (as detailed in ISO/IEC 27001, Chapter 4.3) is defined by the organisation. The organisation that wants its ISMS certified also needs to define the scope of certification (the ISMS scope should not be confused with the scope of certification that encompasses the ISMS scope and any excluded ISO/IEC 27001 requirements). It is the role of the CB to check that this scope is consistent with its business and that it does not exclude anything of their operation, which could affect the organisation's capability, and/or responsibility, to provide information security that meets the security requirements determined by risk assessment and applicable regulatory requirements.

The CB should ensure that the organisation's information security risk assessment properly reflects its activities and extends to the boundaries of its ISMS activities, taking account of any interfaces and dependencies as defined in ISO/IEC 27001 (Chapter 4.3), and it should confirm that this is reflected in the organisation's Statement of Applicability (SoA); see ISO/IEC 27001 Chapter 6.1.3.d. If the ISMS interfaces with services or activities that are not completely within its scope of certification, the risks associated with these interfaces should nevertheless be included in the organisation's information security risk assessment.

10.5.2 Audit Process

The certification of an organisation's information security management system (ISMS) against the standard ISO/IEC 27001 is a two-stage audit process as detailed in ISO/IEC 27006.

10.5.2.1 Audit Stage 1

During this stage, the CB carries out a review of the organisation's ISMS scope of certification and its documentation. This will include consideration of the definition of the ISMS scope and its ISMS policy, the risk assessment report and risk treatment plan, the Statement of Applicability, and other core elements of ISMS. This review enables the auditors to get an understanding and appreciation of the organisation's ISMS. This then will provide a focus for planning the subsequent parts of the audit. It gives an indication of the state of preparedness of the organisation for the audit. Once this review is complete and an audit report is produced, the CB is in a position to decide whether to proceed with stage 2 of the audit. Before proceeding to the next stage, the CB needs to bring together an audit team with the necessary competence to address the auditing of the organisation's ISMS scope of certification.

10.5.2.2 Audit Stage 2

During this stage, the CB audit team visits the site(s) of the organisation to carry out an audit of the ISMS. The objective of this stage is to confirm that the organisation is in comformance with the requirements of ISO/IEC 27001 standard and that it adheres to its own policies and business objectives.

The audit team should clearly focus on the main areas of the standard's mandatory requirements, including:

- The ISMS in the organisational context (ISO/IEC 27001 Clause 4);
- Commitment, leadership and support of the organisation's top management (ISO/IEC 27001 Clauses 5 and 7);
- The risk assessment and risk treatment processes (ISO/IEC 27001 Clause 6);
- The resulting ISMS design and implementation (ISO/IEC 27001 Clause 8);
- The performance evaluation process (ISO/IEC 27001 Clause 9) and the continual improvement (ISO/IEC 27001 Clause 10). The audit team also needs to consider specifics (e.g., what the organisation has put in place for monitoring and measuring ISMS performance and effectiveness, as well as reporting and reviewing against its business objectives and targets).

The assessment will include checking:

- What has been planned and implemented is in operational use to ensure that security;
- That management reviews take place;
- That there is an internal ISMS audit process in place;
- What management responsibility has been defined for implementation and deployment of its information security policy.

The audit team also needs to consider specifics regarding the risk management process: the risk assessment methods used, the risk criteria, the process of determining controls, the SoA and the justification statements in the SoA.

An important aspect of the audit is the presentation and examination of objective evidence to demonstrate the audit trail links between the information security policy, the results of the risk assessments, security objectives and targets, the responsibilities for information security, the system of controls and procedures and what has been put in place for performance monitoring and security reviews.

10.5.3 Nonconformities

In the case of ISO/IEC 27001 a nonconformity is the absence of, or the failure to implement and maintain, one or more requirements that the standards define for an ISMS or a situation that would on the basis of objective evidence raise significant doubt as to the capability of the ISMS to achieve the security policy and objectives of the organisation.

The CB can define grades of deficiency and areas for improvement (e.g., major or minor nonconformities and observations). If such grades of deficiency are used, however they may be defined by the CB, all types of nonconformity should be dealt with as specified in ISO 17021-1 and ISO/IEC 27006.

10.5.4 Audit Report

The audit team is required to provide to the CB management a report of its findings and recommendations as to the conformity of the organisation's ISMS with all of the requirements of the ISO/IEC 27001 standard. Any nonconformity needs to be discharged in order to comply with all of the requirements of the ISO/IEC 27001. These nonconformities should be promptly brought to the organisation's attention and an action plan agreed to resolve them.

The CB shall ask the organisation to comment on the report and to describe the specific corrective actions it will take, or plans it will take within a defined time, to remedy any nonconformity. The closure of such followup actions may need a full or partial reassessment of the ISMS, or a written declaration to be confirmed during surveillance may be considered adequate.

This report should include the identification of ISMS elements audited, the assessed scope of the ISMS being audited, comments on the conformity of the organisation's ISMS with the requirements of the standard with a clear statement of any nonconformity and, where applicable, any useful comparison with the results of previous assessments of the organisation.

The decision whether or not to award a certificate in regard to an organisation's ISMS shall be made by the CB on the basis of the information gathered during the audit, the audit report and any other relevant information. Those who review the report and make the decision to certify shall not have been those that have participated in the ISMS audit to ensure impartiality of the process.

The certificate awarded to the organisation carries the logo of the CB as well as the logo of the accreditation body that has accredited the CB.

10.5.5 Surveillance Audits

The CB will carry out periodic surveillance audits at sufficiently close intervals to verify that the organisation's certified ISMS continues to conform with the requirements of the ISO/IEC 27001 standard, and the ISMS remains effective with regard to achieving the objectives of the organisation's information security policy. This audit should also examine the action taken by the organisation on nonconformities identified during the last audit.

If nonconformities are found during surveillance audits, then these shall be effectively corrected within a time agreed by the CB and to the satisfaction of the CB in accordance to accreditation requirements. If correction of any nonconformity is not made within this agreed time, then the certification shall be suspended or even withdrawn. The time given to carry out corrective action should be consistent with the severity of the nonconformity.

According to the ISMS standard ISO/IEC 27001, the organisation needs to conduct internal ISMS audits at planned intervals to determine whether the ISMS system of controls, processes and procedures conforms to the requirements of the standard. At the same time, the audit should take account of the relevant legislation or regulations requirements and that the ISMS has been effectively implemented, is effectively maintained and performs as expected. An audit programme shall be planned taking into consideration the status and importance of the ISMS processes and areas to be audited, as well as the results of previous audits.

10.5.6 Recertification

Recertification is normally carried out every three years, and its purpose is to verify the overall continuing conformity of the organisation's ISMS to the requirements of the ISO/IEC 27001 standard and that the ISMS has been properly implemented and maintained.

Recertification and surveillance audit programmes should normally include the following:

- Verification that the approved ISMS continues to be implemented.
- Consideration of the implications of changes to the ISMS system of controls initiated as a result of changes in the organisation's operation

and ensuring the overall effectiveness of the ISMS in its entirety in the light of changes in operations. This includes changes to the ISMS documented system.

- Confirmation of continued conformance with the requirements of ISO/IEC 27001 and a demonstration of the commitment to maintain the effectiveness of the ISMS and the effective interaction between all elements of the ISMS.
- Aspects of system maintenance, which are the internal ISMS audit, internal security review, management review, and preventive and corrective action.

The ISMS processes, as specified in the ISO/IEC 27001 standard, provide a management framework for an organisation to realise a programme of continual improvement. This should provide the organisation with effective ISMS implementation and deployment for the common good and governance of its business and which also matches the aims of the recertification and surveillance audit activities mentioned earlier.

10.5.7 Audit Trails

An important aspect of a certification audit is the collection of objective evidence. In gathering such evidence, an auditor commonly looks for a sample of audit trails. For example, the organisation should have carried out a risk assessment, risk treatment and selection of controls exercises leading to a SoA. The auditor should be able to back track this process to find evidence that the organisation has carried out the process correctly. For example:

- The SoA shows a specific control set has been selected for the implementation question, what was the decision making that went behind this selection?
- Tracing backwards, the auditor might then want to discuss this decision-making process in relation to the risk treatment options selected;
- Tracing back further, the auditor might then ask, what risks were identified in the risk report that led to the risk treatment option of reducing the risks?
- Tracing back, the auditor might want to look at the risk assessment report and discuss the control set that has been selected commensurate with the assets, threats and vulnerability they relate to and the associated risks that have been identified.

The auditor will want to check by gathering answers and evidence that there is a correspondence from the controls in the SoA back to the risk assessment that has been carried out. Other examples include checking on the incident handling process that is in place from:

- Detection, reporting and recording the incident;
- To the analysis of the incident;
- To the corrective actions taken to resolve and recover from the incident;
- To the closure of the incident action;
- To followup actions and activities to ensure preventative measures are deployed to avoid reoccurrence of the incident.

Again, the auditor could build up an audit trail of objective evidence that the correct procedures and deployment of information security controls are in place to manage the risks of incident past, present and future. There are many examples in the implementation and deployment of the controls from ISO/IEC 27001 Annex A (i.e., the control statements from ISO/IEC 27002). These might include establishing audit trails relating to:

- Access control measures and procedures;
- Backups;
- Business continuity plans and testing;
- Email and Internet policies;
- Legal and regulatory controls;
- Third-party contracts, SLAs and service delivery management;
- Physical security measures.

Like the two examples given earlier, the auditor should be looking to build up audit trails of objective evidence that the correct procedures and deployment of controls are in place to manage the risks of the business. The auditor will also be looking at the ongoing management of these processes and controls looking for indications and evidence that the organisation is monitoring and reviewing its systems to check how effective its security is and what measures are planned to improve in areas where the security is not so effective.

10.5.8 Competence

Confidence and reliance in the certification audit process is dependent on several things, in particular, the general competence of personnel involved in each of the certification functions and the more specific competence of the auditors conducting ISMS audits. The certification functions include conducting client application reviews, auditing, lead auditor, reviewing audit reports and making certification decisions.

The accreditation requires that the CB has a process in place that defines the necessary competence requirements of its personnel in the certification functions. This includes technological, legal and regulatory knowledge relevant to the client's ISMS being audited.

The CB shall also have a process and criteria in place for verifying the competence of personnel—their experience, training and knowledge. General and specific experience, knowledge and training related to ISMS certifications include the following:

- Management systems and auditing;
- Principles of auditing and the audit process (ISO 19011 and ISO/IEC 27007);
- Management systems principles, philosophy, processes that is ISMS specific;
- General knowledge of information security;
- ISMS standard ISO/IEC 27001;
- Specific processes such as risk assessment and risk treatment (e.g., knowledge of principles and methods defined in ISO/IEC 27005);
- ISMS monitoring, measurement, analysis and evaluation (e.g., knowledge of principles and methods defined in ISO/IEC 27004);
- Information security controls relevant to implementing the ISMS standard ISO/IEC 27001 (e.g., ISO/IEC 27002):
    - Policies and procedures;
    - Human resource security;
    - Access control methods;
    - Operation security;
    - Communications security;
    - Physical and environmental security.
- Technical knowledge of the activity to be audited.

The CB needs to have the competence at selecting auditors with the necessary knowledge, skill and experience to ensure that the audits are carried out effectively and uniformly. The general criteria for auditor competence is based on a demonstration of the ability to apply a range of knowledge and skills gained from formal education, IT and ISMS work experience, ISMS auditor training and auditing experience.

Auditors are expected to develop, maintain and improve their competence through a programme of continual professional development and by participating regularly in ISMS audits.

The International Register of Certificates Auditors has defined a set of criteria for auditor certification based on ISO 19011 (general auditor competence) and ISO/IEC 27006 (ISMS-specific competence). This covers the minimum requirements that auditors need to satisfy in terms of formal education, work experience, auditor training, and auditing experience to be certified as either an ISMS auditor or ISMS lead auditor (see www.irca.org). Also, IRCA has defined the criteria for ISMS training courses based on the ISO/IEC 27000 series of standards. Training organisations can apply to IRCA to have their courses approved according to the criteria. ISO is developing a standard on competence requirements for ISMS professionals (ISO/IEC 27021). This standard is not expected to be published until 2017 at the earliest.

Contents

Epilogos (πλογοσ)

11.1 The ISMS—A Living System

The ISMS should not be treated as a static system. It needs to be treated as a living, dynamic system that will change. If the ISMS is treated as a static system, then after a period of time it will surely become ineffective at protecting the organisation's information assets, it will provide an inadequate quality of protection, and it will be unsuitable to meet the requirements, needs, purpose and objectives of the organisation and interested parties. Such an ISMS will devalue the organisation's investment in the design, development and implementation of its ISMS.

The only way the ISMS can maintain the three objective measures of effectiveness, adequacy and suitability is if the ISMS is treated as a living, dynamic and continual improvement system.

This means that there are a number of critical processes and life cycles that need to continually used, reviewed and refreshed. Most of these have been discussed in the preceding chapters. As a conclusion to this book, we shall present again some (but not all) of these processes in diagram form.

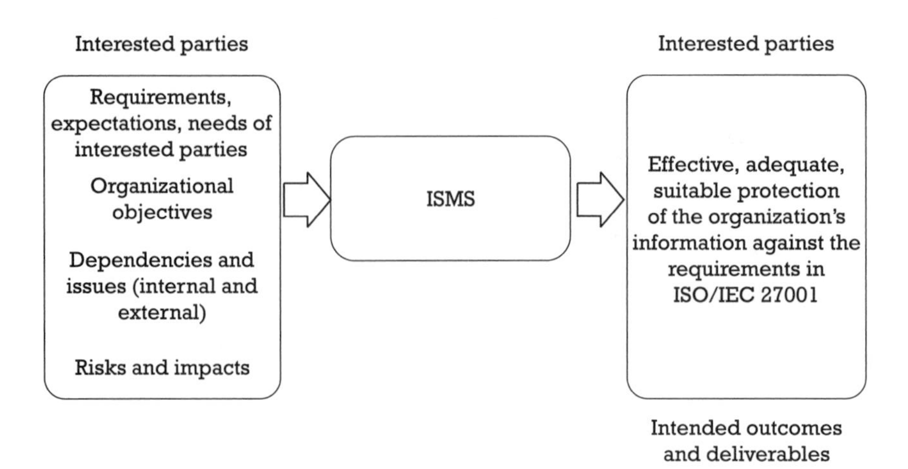

Figure 11.1 ISMS deliverables.

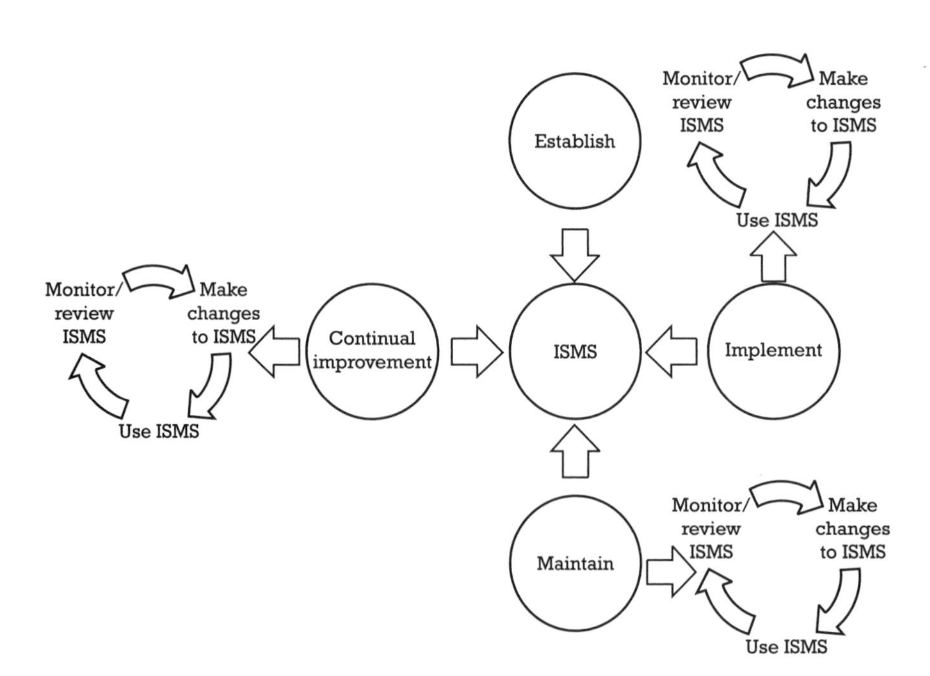

Figure 11.2 Processes, cycles and sequences of activity.

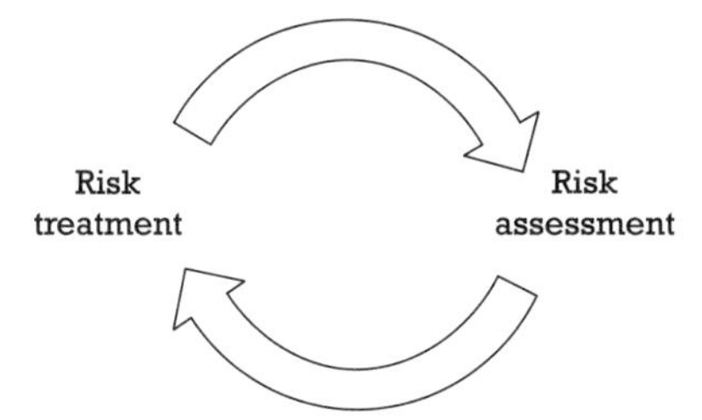

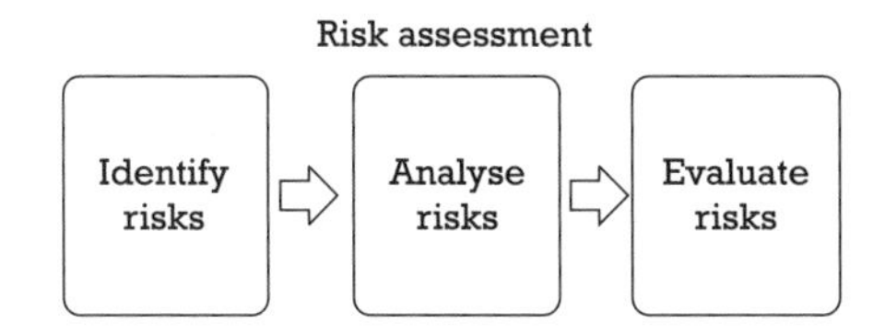

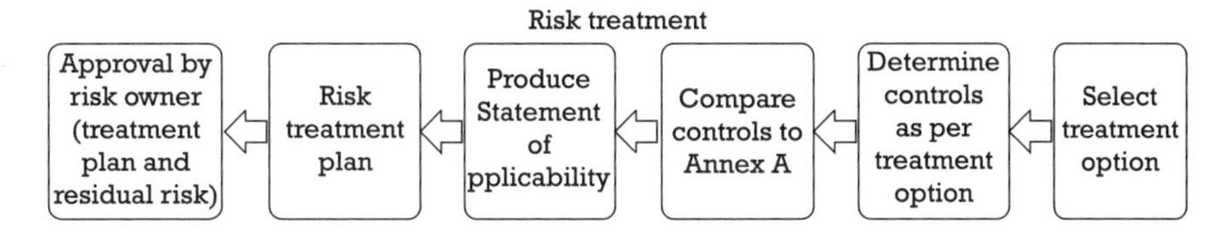

Figure 11.3 Risk management.

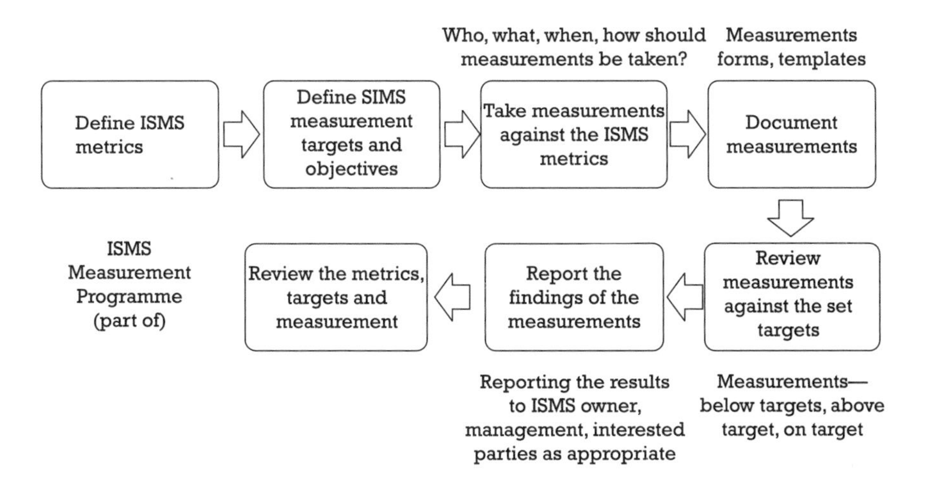

Figure 11.4 ISMS measurements.

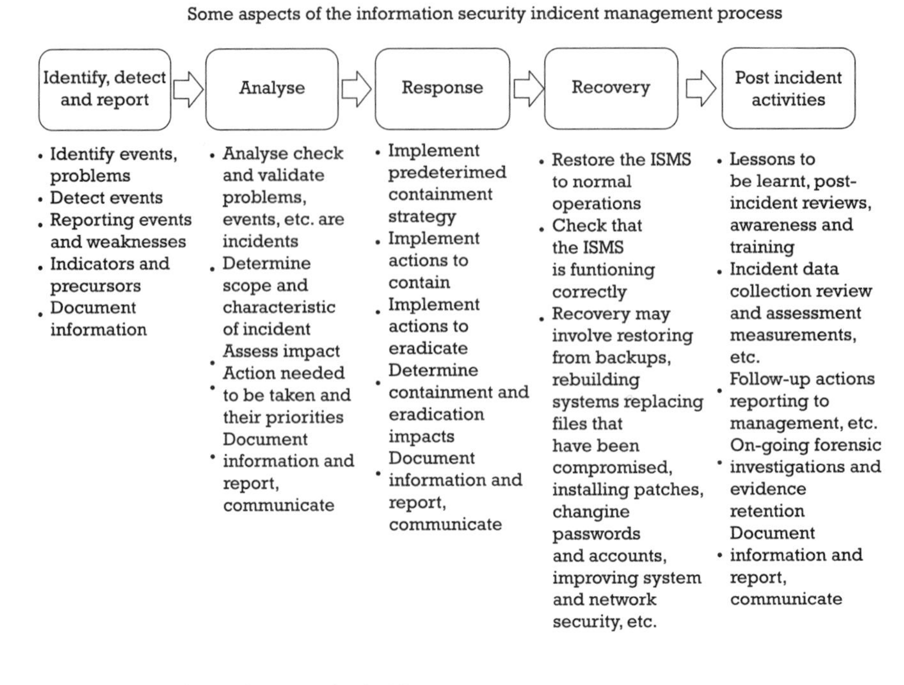

Figure 11.5 Information security incident management.

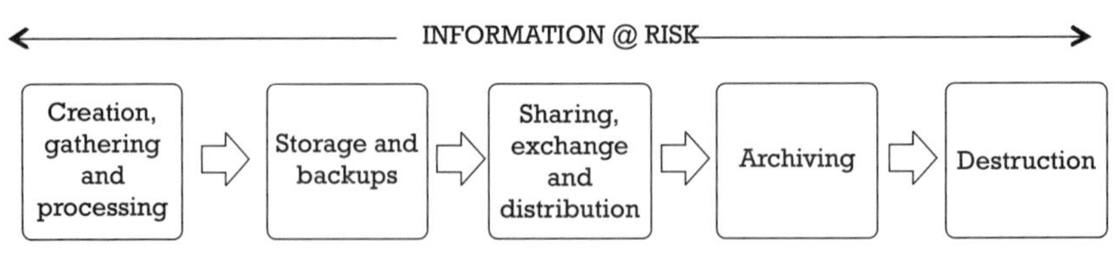

- *Confidentiality* (information is protected against unauthorised access, disclosure)
- *Integrity* (information is protected to ensure its accuracy and completeness, provenance etc., and against any unauthorised modification)
- *Availability* (information is protected to ensure that is available and accessible to those authorised to read and use the information–it has not been denied to authorised people)

- Spoken or oral information
- Information on paper
- Information in electronic form
- Processed manually or using IT
- Single format and types, mixed formats and types, multi-media formats and types, structured and/or structured
- Volume (giga, terra, peta, exa…)
- Gathering, collection, processing at different levels of speed–non-real-time through to real-time, near-real-time, streamed

Figure 11.6 Information life cycle.

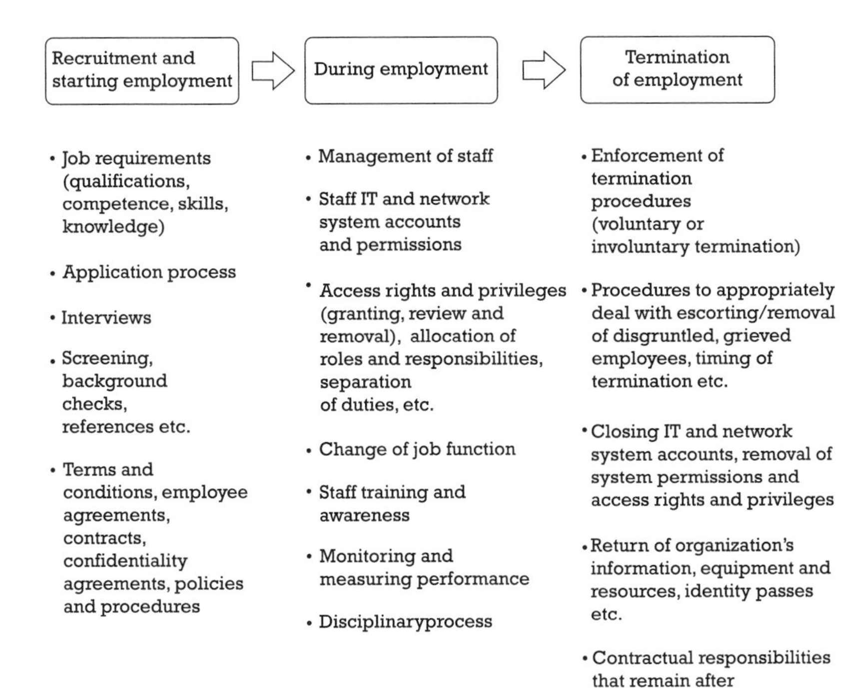

Figure 11.7 Human resources.

11.2 ISMS: The Business Enabler

If ISMS is implemented, used, monitored and reviewed, and undergoes continual improvement to the deliver the intended outcomes against the requirements of ISO/IEC 27001, then the ISMS has value as a business enabler helping the organisation to move forward knowing that its information is effectively, adequately and suitably protected.

- Produce audit plan and schedule
- Discuss the schedule with the client (ISMS owner)
- Select audit team
- Review ISMS documentation
- Request further information documentation

- Implement audit schedule
- Interview ISMS team, users etc.
- Observe, examine, review ISMS being used in practice etc.
- Discuss audit findings with ISMS owner and other interested audit parties, including non-conformities, observations etc.
- Agree deadlines for the resolution of the non-conformities and acceptance of audit findings

- Document the findings of the audit in an audit report
- List non-conformities, observations etc.
- Follow-up-actions
- Deadlines for resolution
- Distribute report to ISMS owner, management and internal audit

- Review audit results by the internal audit department/ group/team
- Review follow-up-actions
- Check the completion of follow-up actions for their satisfactory completion and closured

- Report the findings of an audit finding to the management review meetings
- Discuss opportunities for improvement
- Document discussion and further actions to be taken

Figure 11.8 Internal ISMS audits.

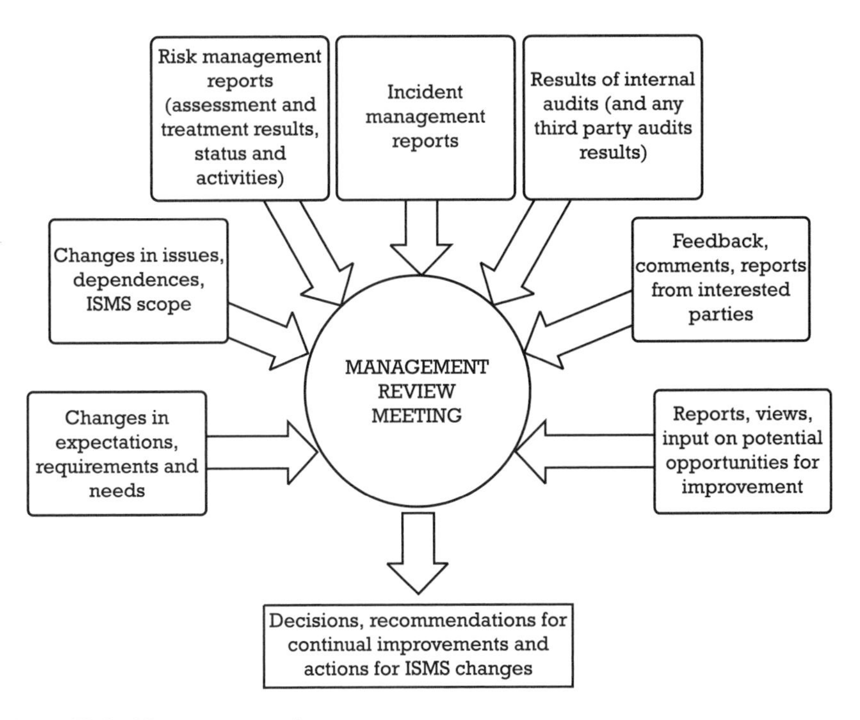

Figure 11.9 Management reviews.

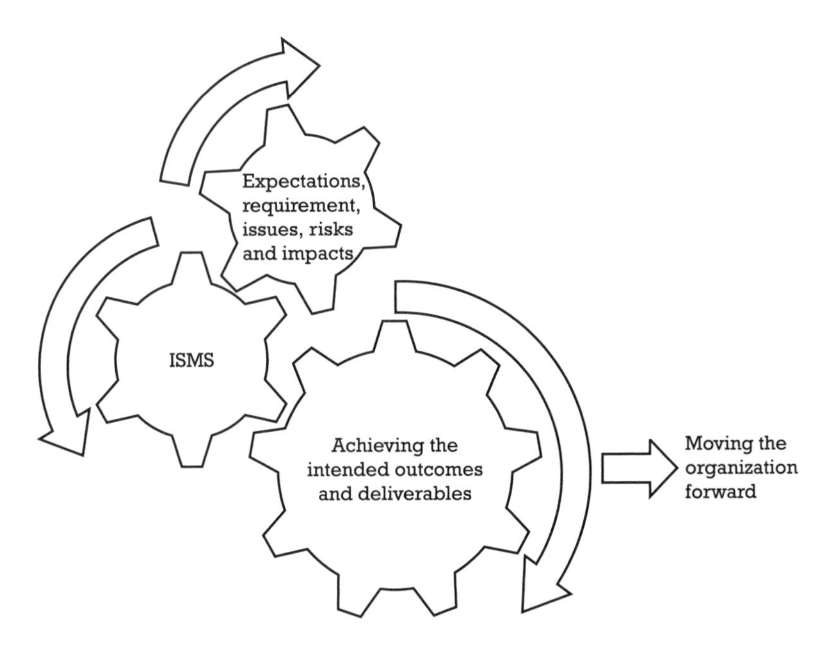

Figure 11.10 ISMS as a management enabler moving the business forward.

Bibliography

[1] BSI BIP 0071:2013, Guidelines on Requirements and Preparation for ISMS Certification Based on ISO/IEC 27001, Edward Humphreys and Bridget Kenyon.

[2] BSI BIP 0072:2013, Are You Ready for an ISMS Audit Based on ISO/IEC 27001? Edward Humphreys and Bridget Kenyon.

[3] BSI BIP 0073:2013, Guide to the Implementation and Auditing of ISMS Controls Based on ISO/IEC 27001, Edward Humphreys and Bridget Kenyon.

[4] BS 7799, Part 1, Code of Practice for Information Security Management, 1995 and 1998 editions.

[5] BS 7799, Part 2, Information Security Management System Risk Management, 2003.

[6] BS 7799, Part 3, Information Security Risk Management, 1998 and 2002 editions.

[7] DTI (UK), Code of Practice for Information Security Management, 1992.

[8] NIST (2013), Special Publication 800-83 Revision 1 Guide to Malware Incident Prevention and Handling for Desktops and Laptops.

[9] NIST (2012), Special Publication 800-80, "Risk Management Guide for Information Technology Systems," Rev. 1.

[10] OECD (2005), "The Promotion of a Culture of Security for Information Systems and Networks in OECD Countries," *OECD Digital Economy Papers*, No. 102, OECD Publishing.

[11] OECD (2012), "The Role of the 2002 Security Guidelines: Towards Cybersecurity for an Open and Interconnected Economy," *OECD Digital Economy Papers*, No. 209, OECD Publishing.

[12] ISO/IEC 27000:2015, Information Technology—Security Techniques—Information Security Management Systems—Overview and Vocabulary.

[13] ISO/IEC 27001:2013, Information Technology—Information Security Management Systems—Requirements.

[14] ISO/IEC 27002:2013, Information Technology—Security Techniques—Code of Practice for Information Security Controls.

[15] ISO/IEC 27003:2010 (under revision), Information Technology—Information Security Management System—Guidance.

[16] ISO/IEC 27004:2009 (under revision), Information Technology—Information Security Management Monitoring, Measurement, Analysis and Evaluation.

[17] ISO/IEC 27005:2011 (under revision), Information Technology—Information Security Risk Management.

[18] ISO/IEC 27006:2015, Information Technology—International Accreditation Guidelines for the Accreditation of Bodies Operating Certification/Registration of Information Security Management Systems.

[19] ISO/IEC 27007:2011 (under revision), Information Technology—Guidelines for Information Security Management Systems Auditing.

[20] ISO/IEC 27013:2015, Information Technology—Guidelines on the Integrated Implementation of ISO/IEC 27001 and ISO/IEC 20000-1.

[21] ISO/IEC 27015:2012, Information Technology—Information Security Management Guidelines for Financial Services.

[22] ISO/IEC 27018:2014, Information Technology—Code of Practice for PII Protection in Public Clouds Acting as PII Processors.

[23] ISO/IEC 27019:2013, Information Technology—Information Security Management Guidelines Based on ISO/IEC 27002 for Process Control Systems Specific to the Energy Utility Industry.

[24] ISO/IEC 27031:2011 (under revision), Information Technology—Guidelines for ICT Readiness for Business Continuity.

[25] ISO/IEC 27035:2011 (under revision), Information Technology—Information Security Incident Management (three parts).

[26] ISO/IEC 29100: 2011, Information Technology—Privacy Framework.

[27] ISO/IEC 29190: 2015, Information Technology—Privacy Capability Assessment Model.

[28] ITU-T X.1031 | ISO/IEC 27017:2016, Information Technology—Guidelines on Information Security Controls for the Use of Cloud Computing Services Based on ISO/IEC 27002.

[29] ITU-T X.1051 | ISO/IEC 27011:2013 (under revision), Information Technology—Information Security Management Guidelines for Telecommunications Organizations Based on ISO/IEC 27002.

About the Author

Professor Edward Humphreys is a philosopher, mathematician and information security management guru. He has work in the field of information security, ICT security and risk management for more than 41 years. During this time, he has been an advisor for major international companies and organisations around the globe, as well as for several European institutions and governments. He is also a visiting professor of ISMS, cyber resilience, risk management and risk psychology, attached to several universities in Europe and Asia.

He is the convenor of the ISO/IEC JTC 1/SC 27 responsible for the development and maintenance of the family of ISO/IEC 27001 ISMS standards. He is often quoted as "the father of ISO/IEC 27001 and ISO/IEC 27002" for his original and pioneering work on the early UK versions of these standards, as well as his seminal work and leadership in the progression of these UK standards into world-best-selling ISO standards. His is also recognised for his work in championing the development and promotion of ISO/IEC 27001 certification around the world. During this formative time, he has been at the forefront in building the international certification framework—from the development of the original accreditation requirements documents for ISMS CBs, through the support of organisations and the development of CBs, acting as a lead assessor and senior advisor for several national accreditation bodies, developing the original ISMS competence requirements for auditors and providing professional training on certification for end users, CB auditors and AB assessors. In addition, he is a leading authority on applied information security risk and incident management, and cyber resilience, in particular, related to the challenges of today's digital world.

In 2002 he was honoured with the Secure Computing Lifetime Achievement Award for his achievements on the internationalisation of information security standards, in particular, as the recognised "father of the ISO/IEC 27001 family of information security management systems standards." In 2014 he was awarded the prestigious UK Wolfe-Barry gold medal award for his outstanding contributions and services to international standards.

Index